P9-DNF-000

THE CAUSEWAY COAST

A GEOLOGICAL EXCURSION GUIDE

CONTENTS

FOREWORD

The spectacular scenery of the North Antrim coast draws many thousands of visitors every year. Indeed it is one of the most famous scenic coastlines anywhere in the world. Like most coastlines frequented by tourists, it has its share of sea, sun and sand. But its really outstanding interests lie elsewhere, in its historical associations and wildlife, and above all in its rocks.

Any landscape is the product of the rocks which underlie it, the forces of nature which have sculpted them and clothed them with vegetation, and the influence of mankind on the land surface over the generations.

The Giant's Causeway is the best known geological feature of the North Antrim coast, but there are many others of considerable interest. This book explains the geology of the area in a global context. It relates how the rocks which now make up Ireland drifted across the surface of the world through geological time, and how they came to shape the extraordinary landscape of this very special area. The Giant's Causeway is now recognised as a World Heritage Site by UNESCO, an accolade given to only one other natural site in the whole of the United Kingdom.

The book also includes illustrated excursion guides, and is written to encourage visitors to delve into and interpret geological history for themselves. Try it. You will not be disappointed.

Dr.John Faulkner
Director of Natural Heritage, Environment and Heritage Service
Department of the Environment (NI)

ACKNOWLEDGEMENTS

I am happy to acknowledge the many people who have helped me in the production of this book. Ian Enlander of the Environment and Heritage Service, DOE has been involved in the project from its inception and his contribution, particularly in the editing process, has been considerable. I am grateful to Kilian McDaid who drew the diagrams and Jennifer Larkin who deciphered and typed the manuscript. I have drawn at length from the maps and publications of the Geological Survey of Northern Ireland, especially The Geology of the Causeway Coast by H E Wilson and P I Manning, which provides an excellent account of the geological history of the area. My thanks to those who read earlier versions of the text and who were willing to try out the Excursion Guides, particularly Edward and Isa Ferguson, John Fulton, Rachel Lyle, John Preston and Joanne Wallace. I am indebted to Marshall McCabe who did his best to enlighten me about the Ice Age and its effects on the landscape. Sylvia Lyle read several drafts in detail and her comments helped to clarify the text in a number of crucial areas. As always her support and encouragement have been invaluable.

Finally I would like to take this opportunity to acknowledge the contribution to the Tertiary igneous geology of Ireland made by Dr John Preston. As a supervisor and colleague he has done much to encourage and stimulate my research interests in the volcanology of the Antrim basalts and time spent in the field with him has always been both enjoyable and productive. Many of the observations and interpretations recorded in this book came from studies carried out with him in north Antrim over more than twenty years, although that is not to say he necessarily agrees with all that I have written.

PREFACE

The Giant's Causeway and the Causeway Coast have long been an attraction for geologists at every level from research scientists to undergraduates to interested lay persons. In addition the superlative scenery has brought visitors with no geological background who have been fascinated and intrigued by some of the most spectacular geological formations to be seen anywhere in the world. The significance of the Giant's Causeway to the development of geology as a science, and its role in the bitter scientific controversies of the late 18th and early 19th centuries, was recognised when the site was declared a World Heritage Site by the United Nations Educational, Scientific and Cultural Organization (UNESCO) in 1986. The Giant's Causeway is an integral part of the geology of north Antrim and the environmental importance of the whole region has been acknowledged by the accordance of the status of Area of Outstanding Natural Beauty (AONB) by the Department of Environment (DOE).

This book is the latest in the Environment and Heritage Service's Interpretation Series commissioned by the DOE(N.I.) to provide information about the landscape, wildlife and heritage of Northern Ireland. The geology of north Antrim is a part of the geology of Ireland as a whole and a full appreciation of the processes involved in the geological evolution of the area can only come from viewing these processes in a global context. To this end the book seeks to explain the geological history of Ireland within the theory of plate tectonics, which is nowadays used to provide a universal framework for earth history. The book provides a résumé of the main points of plate tectonics and the main stages of Ireland's geological evolution and development. The interacting geological forces and processes that have produced the landscape we see today are described in detail. Included are the results of recent research into the eruption conditions that may have influenced the formation of the characteristic columnar basalts of the Causeway.

The final section of the book consists of illustrated excursion guides to the principal geological localities in the area. These guides should allow interested persons to examine for themselves the evidence which geologists have used to piece together the geological history of the area. The text is supplemented by maps, diagrams and photographs and includes a Glossary of technical terms. It is hoped that with the help of this book all visitors, irrespective of their previous geological knowledge, will be able to enjoy the area at a deeper level of understanding than would otherwise be the case.

Biographical note : The author, Dr Paul Lyle, is a Chartered Geologist and Fellow of the Geological Society. He currently lectures at the University of Ulster at Jordanstown.

PHOTOGRAPH CREDITS.

I am indebted to a number of sources for permission to include their photographs in this text. The cover illustration and Figures 3 are used by permission of the Ulster Museum. Figure 17 is by courtesy of Solarfilma, Reykjavik, Iceland. Gordon McDowell took the photographs in Figures 23, 48 and 51 and Peter Millar of the Belfast Geologists' Society took those for Figures 39, 42 and 73. The Geological Survey of Northern Ireland are thanked for permission to use Figure 79.

1 *The Causeway Coast*

1. The Causeway Coast

For over 300 years the spectacular rock formations of the Giant's Causeway area on the north Antrim coast have attracted visitors and scientists from all over the world. In the early 18th century the Causeway itself was the subject of a great deal of controversy concerning the origins of volcanic rocks. It is now known internationally for its contribution to the growth of geology as a science. While earth scientists are interested in explaining the origins of these rocks, other visitors may simply wish to enjoy their scenic value. It is hoped that this publication will go some way towards bridging the gap between the two groups.

1.1 The Area of Outstanding Natural Beauty (AONB).

The Giant's Causeway has been granted the status of World Heritage Site by the United Nations Educational, Scientific and Cultural Organisation (UNESCO), fulfilling the criteria required in representing a major stage of the earth's evolutionary history and containing superlative natural phenomena and features. The Causeway area is owned and managed for the nation by the National Trust and as part of its conservation policy the Department of the Environment (N.I.) has designated the Causeway Coast area an Area of Outstanding Natural Beauty (AONB). The AONB extends from Ramore Head at Portrush eastwards along the north Antrim coast to near Ballycastle. The boundary of

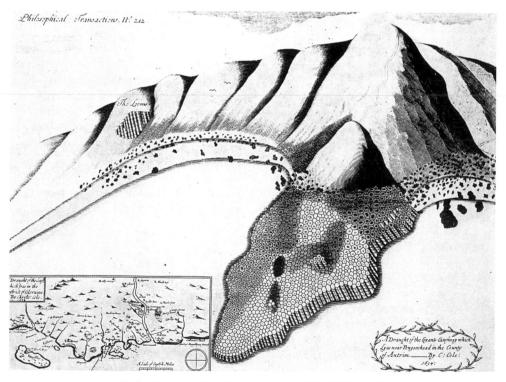

Fig 2 A Draught of the Giant's Causeway 1694. P.T.R.S. 1694

2

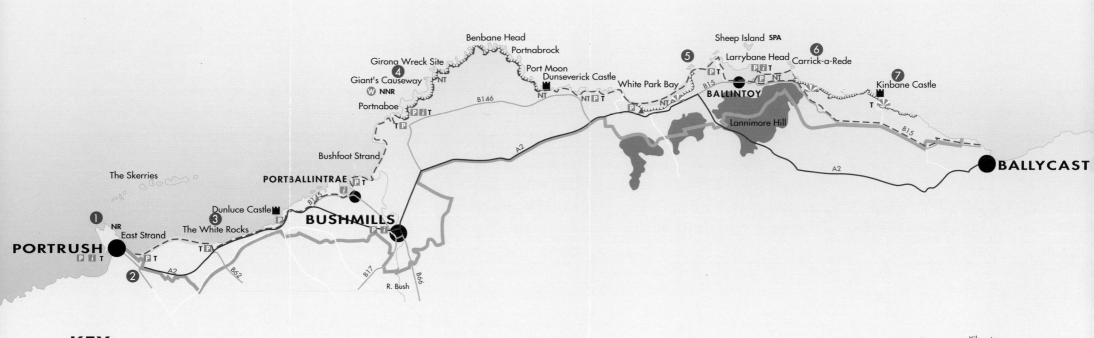

KEY

▬ **AONB Boundary**	– – **Causeway Coast Path**	**NT** National Trust	① Portrush
▬ **Land over 150m**	— River	⚓ Viewpoint	② Craignahullier
▬ **Beach**	Ⓦ World Heritage Site	🅿 Car park	③ White Rocks
— A Roads	**NNR** National Nature Reserve	ℹ Information Centre	④ Giant's Causeway
— B Roads	**NR** Nature Reserve	**T** Toilets	⑤ Ballintoy - White Park Bay
C Roads	**SPA** Special Protection Area	▲ Youth Hostel	⑥ Carrickarede
	▦ Castle		⑦ Kinbane Castle

N

Kilometres

Miles

Map based on O.S. Discoverer Series sheets 4 and 5. Permit No.682 with the permission of the Controller of Her Majesty's Stationery Office, Crown copyright reserved.

Fig 1 Map showing the extent of the Causeway Coast AONB and the excursion localities described in the text.

Fig 3 The west prospect of the Giant's Causeway by Susanna Drury, ca 1740.

the AONB is shown in Fig. 1. Part of the Causeway Coast is also a National Nature Reserve.

This area includes not just the columnar basalts of the Giant's Causeway area but a wide range of other rock types and geological phenomena which are well exposed along this coast. Together they have played an important part in the growth of understanding of earth history both in this country and further afield.

1.2 Susanna Drury, landscape artist.

Although people living in the area from earliest times must have been aware of the existence of the rock structures of the Causeway, it was the end of the 17th Century before it was first referred to in print and it was not marked on a map until 1714. The area was very isolated and up to the middle of the 18th Century the various reports and descriptions from occasional visitors meant that only rather confused accounts and drawings of the Causeway had been produced. The earliest known illustration of the Causeway is an engraving of a crude aerial view by Christopher Cole (Fig. 2) published in the Philosophical Transactions of the Royal Society in 1694. While this gives a good impression of the hexagonal nature of the joints, any other geological information is vague and confused.

In 1740 an unknown Dublin artist Susanna Drury, after spending 3 months in the area, produced paintings of the East Prospect and West Prospect of the Giant's Causeway showing in remarkably accurate detail, not only the structure of the basalt columns, but also the cliffs and bays around the Causeway (Fig. 3 and front cover). These paintings were engraved by

Francois Vivares, a Frenchman who worked in London, and prints produced from these engravings were widely circulated in scientific circles in Europe. These prints were to play a major role in a bitter scientific controversy that was then raging. The Giant's Causeway is composed of a rock known as BASALT, which is an IGNEOUS ROCK, that is it forms by the cooling and hardening of molten material produced from deep in the earth at temperatures of around 1100 degrees C. Throughout the 18th century however nat-

ural scientists had been debating the origin of crystalline rocks such as basalt and two principal schools of thought had developed. The "NEPTUNISTS" led by Abraham Werner (1749-1817) regarded all such rocks as the result of crystallisation from sea water, while the "VULCANISTS", with Nicholas Desmarest (1725-1815) as their main spokesman, believed them to be the products of volcanic eruptions.

Using the evidence from the Vivares engraving, Desmarest was able to compare the jointing in basalt visible at the Giant's

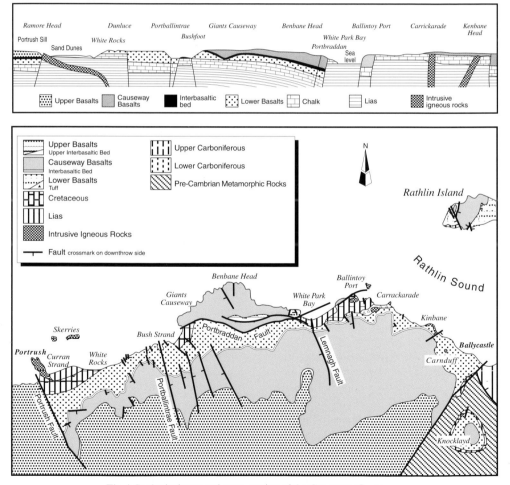

Fig 4 Geological map and cross-section of the Causeway Coast area
(after Wilson and Manning, 1978).

4

PERIOD	AGE(MY)	IRISH ROCK TYPES	GEOLOGICAL ENVIRONMENT	PLATE PROCESSES	EVOLUTION OF PLANT AND ANIMAL LIFE
QUATERNARY — Holocene		Blown sand, Peat, Sands, clays.	Beach lake and river deposits.	Continued widening of the north Atlantic.	Development and dominance of modern toolmaking man. Rise and extinction of woolly mammoth and rhinoceros.
QUATERNARY — Pleistocene	2	Sands, gravels, Boulder clay.	Glacial deposits from ice-caps and glaciers.		Neanderthal man.
TERTIARY — Pliocene	7	Clays, lignites (Lough Neagh).	Marsh and lake deposits.	Himalayas formed by collision of India with Europe.	Early men in Africa and widespread occurrence of grazing and carnivorous mammals.
TERTIARY — Miocene	26				First primitive apes.
TERTIARY — Oligocene	38				
TERTIARY — Eocene	63	Basalt lavas.	Fissure eruptions.	North America splits from Europe and north Atlantic opens.	Whales and dolphins evolve. Main bird groups now present.
TERTIARY — Paleocene	65				
CRETACEOUS	145	White limestone with flints.	Marine conditions with the accumulation of plankton skeletons.	South Atlantic widens, Ireland in the position now occupied by Spain.	Extinction of dinosaurs, plesiosaurs and ammonites. First birds evolved. Early flowering plants.
JURASSIC		Mudstone and thin limestones with fossils.	Marine conditions.	Breakup of Gondwanaland continues.	Dinosaurs dominant on land with plesiosaurs in the oceans and pterosaurs in the air. Early mammals.
JURASSIC — Lias	215	Dark shales and mudstones.	Coastal lagoons with rising sea levels.	Super-continent Gondwanaland splits and south Atlantic opens.	First dinosaurs and large marine reptiles. First flies appear. Ammonites common in the oceans.
TRIAS	250	Sandstones and salt deposits.	Shallow seas, with seawater evaporation. Shallow waters in arid continental conditions.		
PERMIAN	285	Limestones, sandstones and breccias.	Desert conditions with occasional marine periods.	Ireland in desert belt north of Equator.	Great spread of reptiles on land. Insects continue to diversify and spread. Conifers common.
CARBONIFEROUS	360	Limestones, sandstones and thin coals.	Marine conditions with reef-forming corals. Coal seams formed in coastal lagoons.	Ireland on Equator.	Spread of amphibians and shark-like fishes. Insect evolving rapidly. Early trees abundant. First reptiles appear.
DEVONIAN	410	Conglomerates, sandstones and mudstones.	Continental deposits in arid conditions with periodic floods.	Ireland in desert belt south of Equator.	First amphibians. Fern-like forms on land.
SILURIAN	440	Greywackes, shales and sandstones.	Mainly marine and formed in deep water by submarine slumps of material off the continental shelf to deep water. Occasional volcanic activity.	Mountain building period and Iapetus Ocean closes.	First land plants appear. Armoured jawless fishes common.
ORDOVICIAN	505	Occasional lavas and volcanic ashes.			Early fishes appear. Graptolites, trilobites and brachiopods abundant in the oceans.
CAMBRIAN	590	No examples in Northern Ireland. Probably lost by erosion.		Early super-continent breaks up and proto-Atlantic or Iapetus Ocean opens.	Dominance of trilobites in the shallow seas and development of early shelled forms such as brachiopods.
PRE-CAMBRIAN	1500	Crystalline metamorphic rocks: schists, gneisses and marbles.	Sedimentary rocks changed by later earth movements and mountain building processes.	Beginning of Plate Tectonic processes after formation of Earth's core, mantle and crust.	Early multicelled animals including jellyfish-like forms and worm-like animals.
PRE-CAMBRIAN	4600	Age of Earth			Early bacteria and algae

Fig 5 Stratigraphical column showing the order of succession of the main rock types found in Ireland and their conditions of formation.

Causeway with similar structures he had observed in the Auvergne district of central France which were clearly the result of LAVA flows issuing from an obvious volcanic crater. By the early part of the 19th Century the controversy between Neptunists and Volcanists had subsided, and the volcanic origin of igneous rocks had been generally accepted. The Giant's Causeway was by then recognised as an important site in the history of geology, attracting scientists from all over the world. In addition the area was becoming popular with other visitors and the continued development of nearby Portrush as a Victorian seaside resort meant that growing numbers of visitors made their way to an increasingly commercialised Causeway. Travel by jaunting car or boat and later by the worlds first hydro-electric tramway made the locality more easily accessible. The Causeway still acts as a magnet for visitors and is currently the single most popular tourist attraction in Northern Ireland, while at the same time the subject of continuing scientific research. It is the destination for many visiting geologists from all over the world who come to see a classic locality that has been referred to in so many geology text books published over the last 100 years or more.

1.3 The geology of the Causeway Coast: an introduction.

While the best known geological feature of the Causeway Coast AONB is the columnar basalts of the Giant's Causeway itself, there are other rock types in the area and to fully appreciate the significance of the Causeway it is necessary to look at the overall geological history of the north Antrim coast. The geological map, Fig. 4, shows the distribution of the main rock types in the area. The order in which the rocks occur in the area is shown by geologists using a STRATIGRAPHICAL COLUMN such as Fig. 5 .The oldest rocks are at the bottom of the column or succession and become progressively younger upwards, with the youngest FORMATIONS at the top.

Geological time has been divided into a number of PERIODS and the most recent of these, the QUATERNARY, began 2 million years ago. This period can be subdivided into the PLEISTOCENE, which was the time of the last great Ice Age, and the HOLOCENE which continues up to the present day. Rocks of Quaternary age in the area are mostly the boulder clays and glacial sands and gravels left by the various ice caps and glaciers from up to 17000 years ago and the more recent peat and wind blown sands. These types of deposits are known as DRIFT and are not marked on the geological map which shows only the occurrence of the underlying bedrock or SOLID geology.

The rocks which lie directly under the drift deposits, and are therefore older, are the basalts of the Antrim Lava Group, including those of the Giant's Causeway, which are TERTIARY in age. The Tertiary period lasted from about 65 million years ago until 2 million years ago, and the basalts began to erupt around 60 million years ago. Below the Tertiary lavas are the deposits of white LIMESTONE or Chalk which formed in the CRETACEOUS period which lasted from 135 million years to 65 million years ago. The oldest rocks exposed along the Causeway Coast are the LIAS clays and MUDSTONES which are part of the JURASSIC period which lasted from 190 million years to 135 million years ago. Elsewhere in Antrim it is these soft clays which are responsible for the mud flows and landslips that occur on the

Coast Road, but in north Antrim they appear mainly in the White Park Bay area and in a modified form (to be discussed later) around Ramore Head at Portrush.

The Jurassic and Cretaceous in western Europe were periods when the region was covered by the sea. The Jurassic sea was fairly shallow and the resultant Lias deposits are mostly mudstones and clays with thin limestone beds, containing abundant fossil remains. Fossils recovered commonly from the Lias clays include marine invertebrates such as sea urchins, AMMONITES, an extinct group which were related to modern NAUTILOIDS, and also various marine reptiles such as PLESIOSAURS. By Cretaceous times, a warm shallow sea covered all of western Europe, including Britain and Ireland, and stretched as far east as Russia. On the bottom of this sea the white Chalk was forming from the accumulated skeletons of floating micro-organisms which had lived in the sea. Through time these skeletons built up thick deposits of lime-rich mud which eventually hardened to form the characteristic white Chalk seen around the Antrim coastline. By about 70 million years ago the sea level had dropped all over Europe and north-east Ireland was

again a land surface. By the beginning of the Tertiary, about 65 million years ago, Antrim was a rolling landscape of low hills with pockets of vegetation. This was prior to the period of continental splitting or RIFTING which was to be accompanied by the long period of volcanic activity which produced the Antrim Lava Group. The Antrim Lavas were erupted in two main phases giving the Lower Basalt Formation (the Lower Basalts) and the Upper Basalt Formation (the Upper Basalts), with a long spell of relative volcanic dormancy during which the rocks of the Interbasaltic Formation were produced. The columnar basalts of the Giant's Causeway, the Causeway THOLEIITE Formation, occur as part of the Interbasaltic Formation. This rifting and break-up of continents has taken place all over the Earth throughout geological time, and is a process that is continuing today. To understand fully the processes that were responsible for the Antrim basalts it is necessary to look at those forces that have modified the earth's surface throughout approximately 4,500 million years of geological time and to examine in greater detail the geological history of Ireland.

2
Wandering Continents

2. Wandering Continents

2.1 Plate Tectonics.

In the roughly five thousand million years since the formation of the earth from the swirling cloud of gas and dust that marked the beginning of the Solar System, the planet's interior and outer surface have undergone many changes, some of them catastrophic in their effects.

A glance at today's world map shows Ireland located in the northern hemisphere, slightly west of the Greenwich meridian, and in the broadly temperate climatic zone, lacking the extremes of temperature shown by the polar zones to the north or the tropical latitudes to the south. There is a natural tendency to feel that this must always have been the case, but examination of some of the rocks found in Ireland shows that they formed under climatic conditions found elsewhere on the surface of the Earth. Probably the most striking examples of this are the limestones in Fermanagh which contain abundant fossil coral reefs that could only have formed in a warm tropical sea, or the red SANDSTONES such as those at Scrabo near Newtownards that must have formed in hot deserts near the Equator in conditions similar to the present day Sahara.

Geologists now recognise that the pattern of the earth's continents and oceans is not fixed or permanent, but that the continents and associated ocean floors move slowly over the surface of the Earth at the rate of a few centimetres per year, with the oceans first opening and expanding, and then gradually, through time, closing again. This process is referred to as CONTINENTAL DRIFT. At present the Atlantic Ocean is widening at the rate of a few centimetres per year which means that north and south America are moving progressively further from Africa and Europe. At the same time the Pacific Ocean, which is older than the Atlantic, is gradually shrinking. Most of the earthquakes and volcanic eruptions of the Earth result from these large scale changes in the continents and oceans, while the energy required to propel the continents across these vast distances comes from heat generated inside the Earth. The crust of the Earth consists of a relatively small number of irregularly shaped segments or PLATES which move slowly in relation to each other and thus produce those changes in the distribution of the oceans and continents seen on the surface throughout geological time. As TECTONICS is the name given to the branch of geology that studies the movements and deformations of the earth's crust, these large scale processes are known as PLATE TECTONICS.

The map of the earth shown in Fig. 6 shows the present position of the continents and oceans together with the boundaries of the main plates and the directions in which they are currently moving. For example the plates labelled AFRICAN and EURASIAN are moving in the opposite direction from those labelled AMERICAN, thus widening the Atlantic Ocean in the process. It can also be seen that some plates, such as the American, consist of ocean and continental areas, while plates such as the Pacific are entirely oceanic.

2.2 Plate Tectonics and Ireland.

These changes have been taking place for all of geological time but most of the oldest rocks have long since been destroyed or buried. However sufficient evidence remains from younger rocks to be able to plot the global wandering of Ireland from around 1000 million years ago to the present.

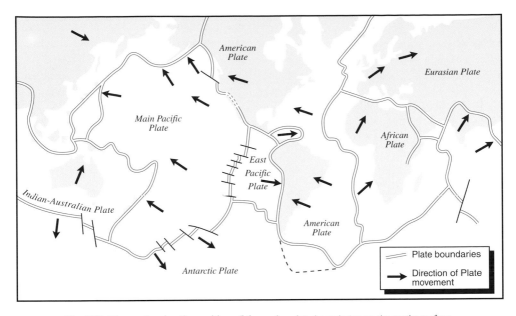

Fig 6 World map showing the position of the major plate boundaries on the earths surface and their directions of movement.

When the oldest rocks now found in Ireland were being formed (about 1000 million years ago in the PRE-CAMBRIAN, see Fig. 5) that segment of the earth's crust in which they were forming and which we now recognise as part of Ireland was in the southern hemisphere. In fact it was about 20 degrees south of the Equator (compared with about 55 degrees north at the moment), and in the position currently occupied by the Indonesia. Fig. 7 shows the approximate position of Ireland 1000 million years ago. Over the following 1000 million years of geological time Ireland was to undergo many environmental changes, sometimes under deep oceans, at other times forming exposed land areas, and all the time accumulating rock types representative of these varying geological conditions, clues which have allowed geologists to unravel its complex geological history.

For a reconstruction in time and space of the movement of these continental landmasses geologists and other scientists interpret information that exists in the physical and chemical properties of many rocks. It is possible to calculate the age of a rock by measuring the amounts of certain radioactive elements that are present in very small quantities in their minerals. Uranium, for example, is unstable and decays over time giving out energy in the form of radiation and the rate at which the energy is emitted is constant over very long periods. This process is called RADIOACTIVE DECAY and when an element such as uranium decays and produces radioactive energy, it can change its atomic structure and becomes another element. One of the elements produced by the radioactive decay of uranium for instance is the element lead. By analysing the amounts of uranium and its decay product lead in the rock, and by measuring in the laboratory the rate at which this decay occurs, it is possible to calculate how long the process has been going on - that is

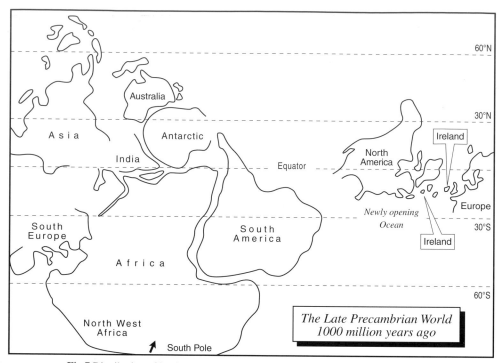

Fig 7 Distribution of landmasses and the position of Ireland in the late Pre-Cambrian, 1000 million years ago, after Smith et al, 1973.

when the mineral crystallised, which was when the rock was formed. This is called RADIOMETRIC DATING and by carrying out this calculation for very many samples it has been possible to construct a timescale for the rocks of the earth (Fig. 5).

Radiometric dating can give information on the age of the rock but tells nothing about its location on the surface of the Earth when it was first formed. To work out this original position, information from the earth's magnetic field is required. The earth's magnetic field, or GEOMAGNETIC FIELD can be measured at any point on the surface of the Earth. Most people have used or watched someone else use a compass and the end of the compass needle marked north gives the direction of the MAGNETIC NORTH POLE from that point. It is less commonly known that the same compass needle, as well as swinging

round to magnetic north, is also pointing down into the Earth, and the nearer the magnetic pole the compass needle is, the steeper the angle at which the needle dips. This feature of the geomagnetic field can be used therefore to calculate how far away the magnetic pole is from that point, in the direction indicated by the arrow. Rocks such as basalt contain small crystals of an iron ore mineral known as MAGNETITE, also called LODESTONE, which the ancient Chinese used as a compass as long ago as the first century AD. When a basalt lava has cooled the magnetite crystals take on the imprint of the geomagnetic field of the time and place of their origin and this is retained in the rock as the "fossil" or PALAEO-MAGNETISM, and can be used to work out the position of the magnetic north pole, the PALAEO-MAGNETIC POLE, for the time of formation of the lava.

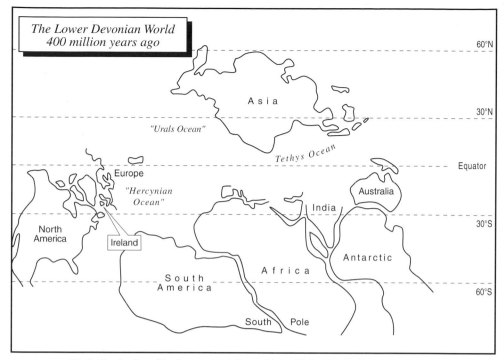

The Lower Devonian World
400 million years ago

60°N

A s i a

30°N

"Urals Ocean"

Tethys Ocean

Equator

Europe

"Hercynian
Ocean"

Australia

India

30°S

North
America

Ireland

Antarctic

Africa

South
America

60°S

South   Pole

Fig 8 Distribution of landmasses and the position of Ireland in the Lower Devonian,
400 million years ago, after Smith et al, 1973.

When this is done for rocks in Ireland which are 500-1000 million years old, the positions of the north pole they indicate mean that Ireland must have been at approximately the latitude of Indonesia and south of the Equator at that time, hence the position marked on Fig. 7.

One other major difference at this stage in the geological history of Ireland was that the northern half of Ireland, along with Scotland, was part of the North American continent, separated by a widening ocean from the southern half of Ireland which was joined to England and Wales and part of Europe. By dividing America from Europe this ocean was an earlier version of the Atlantic and has been named IAPETUS, (the Atlantic Ocean derives its name from Atlas, the figure in Greek mythology who supported the sky on his shoulders, and whose father was Iapetus).

At this time life forms were restricted to the seas and consisted mainly of various types of shell-fish and PLANKTON. Iapetus existed throughout the Ordovician and Silurian Periods (Fig. 5) and by 400 million years ago had opened to its maximum width and then closed again, forming a major mountain chain as the two continents collided. The dark blue/grey sandstones found in most of Counties Down and Armagh were formed in this ocean while the upland areas of Donegal and the Sperrins in Ireland and the Highlands of Scotland are the remains of that mountain chain. By this time Ireland, with the two parts now joined together, had drifted eastwards around the world to near the present position of central America. It was still south of the Equator, and was now part of a large continent consisting of Scandinavia, Greenland and parts of north America as shown in Fig. 8.

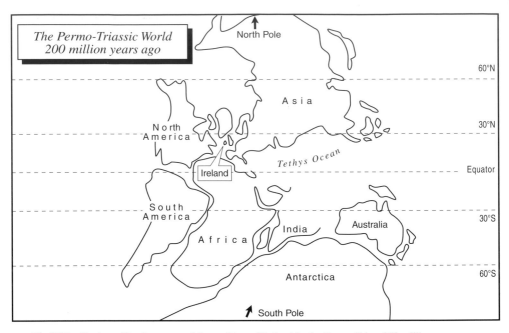

The Permo-Triassic World
200 million years ago

North Pole

60°N

Asia

30°N

North
America

Tethys Ocean

Ireland

Equator

South
America

30°S

India

Australia

Africa

Antarctica

60°S

South Pole

Fig 9 Distribution of landmasses and the position of Ireland in the Permo-Trias, 250 million years ago, after Dietz and Holden, 1970.

This continent had high mountains to the north dominating lowlands to the south. Rocks of this part of the Devonian are known as the OLD RED SAND-STONE and are found in Ireland at Cushendun or in Tyrone for example. They formed in deserts with conditions similar to hot continental deserts such as the Sahara. Primitive fishes now lived in the oceans and the first plants began to colonise the land areas.

For the next 50 million years or so Ireland continued to drift eastwards and also northwards, so that by 350 million years ago Ireland was in tropical latitudes, on or near the Equator. The area was dominated by a generally shallow sea bounded to the north by the greatly reduced continent that had existed during the previous Devonian Period. This was the CAR-BONIFEROUS period (Fig. 5), the sea was rich in coral reefs and the land areas

were covered in lush vegetation with giant trees forming thick forests in which amphibians and early reptiles lived. Around the coasts, tangled mangrove forests built up thick layers of vegetation which eventually decayed to form the extensive coal seams found in rocks of this age over much of Europe. The limestones that are found commonly in Fermanagh formed in these warm seas and contain many examples of the coral reefs that lived there. The coal seams around Coalisland and Ballycastle formed in swampy regions on the edge of these seas.

This phase of luxuriant plant growth ended as Ireland continued its drift northwards across the Equator and back into desert latitudes, but this time north of the Equator during the PERMIAN and TRI-ASSIC periods (Fig. 9). These periods lasted from about 285 to 215 million years ago and by 250 million years ago Ireland

12

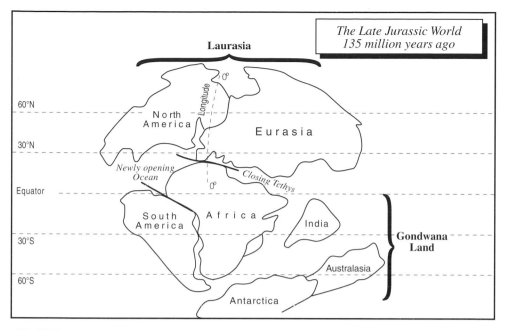

Fig 10 The components of the supercontinent Pangaea, Laurasia and Gondwanaland, 135 million years ago, after Dietz and Holden, 1970.

was near the present position of the Sahara desert and was again experiencing hot dry conditions, with sand dunes, dried river and lake beds and sparse plant life. The reddish sandstones and mudstones of Scrabo Hill at Newtownards were formed at this time while the thick salt deposits mined at Carrickfergus resulted from evaporation of seawater from shallow coastal lagoons by hot desert sun. This period was the beginning of the age of the giant reptiles, with dinosaurs on land from mid-Trias onwards.

Britain and Ireland were by this time part of a huge continental landmass called PANGAEA, made up of all the continents on the earth's surface. This "super-continent" was splitting and the north part (called LAURASIA), consisting of what is now North America, Europe and Asia, was moving away from the southern part (called GONDWANALAND), made up of South America, Africa, India, Australia and Antarctica (Fig. 10).

By 100 million years ago Ireland, still moving northwards, had reached the position now occupied by Spain, and the present Atlantic Ocean was opening, splitting Europe from Greenland and North America (Fig. 13a). At the same time in the southern hemisphere, South America and Africa had drifted apart and Australia was moving away from Antarctica towards its present isolated position.

The hot deserts of earlier times had been replaced by long periods when Ireland was covered by warm shallow seas with occasional periods when sea-levels fell and the land dried out. It was during these submerged episodes that the white Chalk seen on the Antrim Coast Road was formed.

The distribution of the continents 50 million years ago was very similar to what

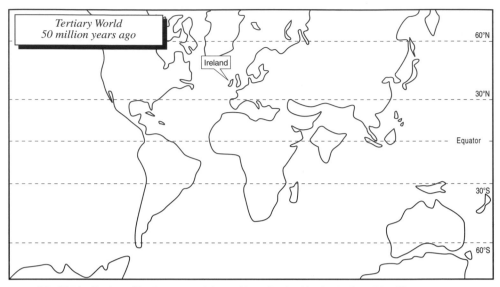

Fig 11 Distribution of landmasses and the position of Ireland in the Tertiary, 50 million years ago, after Smith et al, 1973.

it is today (Fig. 11). Note that the Atlantic was much narrower than at present and India had not yet come into collision with the southern coast of Asia to form the Himalayas. Ireland had travelled as far north as the position of the French Riviera and was part of a major area of volcanic activity formed as Europe and North America were pulled apart to form the North Atlantic (Fig. 13b). The volcanoes and lava flows that now make up the Antrim Plateau were formed during a period of volcanic eruptions that probably lasted several million years. The map of the present distribution of continents and oceans shows the changes in the last 50 million years (Fig. 6). During that time the Atlantic has widened, India collided with Asia and the force of the impact has buckled the crust producing the mountain ranges of the Himalayas. This collision is still going on and the whole Himalayan region continues to rise at the rate of a few centimetres per year. That the Atlantic is still widening is shown by the frequent

volcanic eruptions in Iceland which sits astride the "join" between the two halves of the spreading ocean floor.

In time the Atlantic Ocean will reach its maximum width and begin to close, as it has done before in the past, and as the Pacific is doing at the moment. That will begin the whole mountain-building cycle over again and Ireland will continue its geological odyssey for another few billion years until the internal energy of the Earth becomes insufficient to drive the vast heat engine which has propelled Ireland this far.

2.3 Before the Causeway.

By the end of Cretaceous times (Fig. 5, stratigraphic table), about 65 million years ago, the whole of Ireland, along with most of western Europe, had been submerged under a warm shallow sea in which the white limestone or Chalk that is such a prominent feature of the Antrim coast was deposited. This very characteristic rock type can be found throughout Europe as

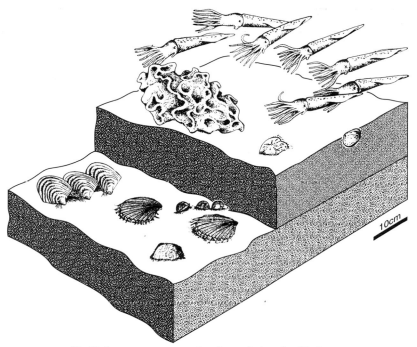

Fig 12 Cretaceous seascape showing typical marine life-forms.

far east as the Caucasus Mountains between the Black Sea and the Caspian Sea in Russia. This extensive sediment is almost entirely derived from the remains of shells of microscopic planktonic life forms called COCCOLITHS. After the death of the organism the calcareous disc contained within the cell fell to the sea bed to form a mud which eventually formed fine grained white chalk. Larger fossils also existed in this sea with remains of bottom-dwelling sea urchins commonly found, along with a large number of remains of squid-like animals called BELEMNITES. Fig. 12 gives an impression of life in a Cretaceous sea. Associated with the Chalk in Antrim is FLINT, usually in the form of rounded nodules and often arranged in relatively thin layers. Flint is a very fine-grained form of the mineral QUARTZ which breaks to give a very characteristic shell-like or CONCHOIDAL FRACTURE. The fracture often produces a very sharp edge and this property was utilised by the earliest inhabitants of the area to produce flint tools such as axes, arrowheads and scrapers. The presence of flint in such quantities was an important factor in the settlement pattern of north-east Ireland some 9,000 years ago.

By the beginning of the Tertiary Period (Fig. 5, stratigraphic table) Ireland had emerged from this warm sea and a low rolling land surface had developed after several million years of weathering and erosion. This weathering produced a layer of red-brown clay soil found frequently on the upper surface of the Chalk in Antrim, and referred to as "clay-with-flints". This was the land surface on to which the lavas of the Antrim plateau were erupted, starting about 65 million years ago. The timing and the location of this volcanic activity were not random but closely associated

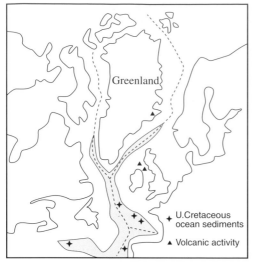

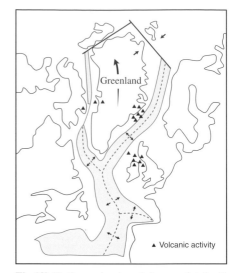

Fig 13a The opening of the north Atlantic in Upper Cretaceous times.

Fig 13b Tertiary volcanic activity associated with the opening of the north Atlantic, after Harland, 1969.

with processes operating on a much larger scale in the northern hemisphere, i.e. the opening and widening of the North Atlantic Ocean and the separation of the continental blocks of North America and Europe (Fig. 13a).

The late Cretaceous world was one in which a southern Atlantic ocean divided South America from Africa, India was a sub-continental island and there existed a vast northern continent containing north America, Europe and Asia. It was the break up of this immense continental land-mass in Tertiary times which provided the focus of volcanic activity in the northern hemisphere starting about 65 million years ago (Figs. 13a and b).

2.4 The Opening of the North Atlantic Ocean.

By the start of the Tertiary Period around 65 million years ago, Ireland was situated on the Earth's surface at about the latitude presently occupied by northern Spain and the south of France. The South

Atlantic was open and widening and spreading activity began to move north-wards with the main focus in the area between Greenland and Britain and Scandinavia. Rifting associated with the northward growth of the Atlantic was responsible for the volcanic activity that occurred extensively in north-east Ireland and also north-west Scotland around this time.

2.5 The Causes of Plate Motion.

Earth scientists now have a reasonably clear picture of the movement of the plates on the earth's surface in terms of the directions and rates of progress as already detailed for Ireland, however there is still no overall explanation of the precise caus-es of plate motion. The Earth can be subdi-vided into concentric layers. The outer rel-atively thin and brittle layer is the CRUST, this overlies a thick layer going down to almost 3000 kms from the surface, the MANTLE, which although a solid is hot and capable of flowing very slowly like a liquid. The central zone is the CORE consisting predominantly of iron with a liquid outer part and a solid inner (Fig. 14).

The plates are formed of the continental or oceanic crust with the brittle upper part of the mantle and these regions together are known as the LITHOSPHERE. The region in the mantle directly beneath the lithosphere is called the ASTHENOSPHERE and is partially liquid due to mantle melting. It may be this layer along which the tectonic plates move or "drift" (Fig. 14).

The precise causes of plate movement are likely to be complex and involving a number of factors but it is probable that the process of MANTLE CONVECTION is at least a major part of the mechanism.

CONVECTION is the movement of material as a result of temperature differences. For example a saucepan of porridge being heated on a hob will have a set of convection currents set up within it. Heat from the cooker enters the porridge at the bottom of the pan and is carried upwards by the heated porridge, which rises because it is less dense, or lighter, than the colder unheated porridge. This hotter material however cools at the surface of the pan and thus becomes heavier and then sinks to the bottom of the pan, to be recycled as part of a CONVECTION CELL when it returns to the surface of the pan after re-heating (Fig. 15a). The same process is thought to operate within the mantle where the heat source is from radioactive materials deep in the earth's interior (Fig. 15b).

In simple terms the rising parts of convection cells occur underneath those parts of the crust where the plates are moving apart, and since the crust in these regions is being stretched and new crust is being formed, they are referred to as CONSTRUCTIVE PLATE MARGINS. Many of these constructive plate margins are now in the oceans and occur as the vast submarine mountain ranges known as the MID-OCEAN RIDGES (Fig. 15c). The sinking limbs of the convection cells occur where plates are moving together, such as around the rim of the Pacific, and since the crust here is being compressed and shortened by one plate being forced down below the other, (Fig. 15c), they are described as DESTRUCTIVE PLATE MARGINS.

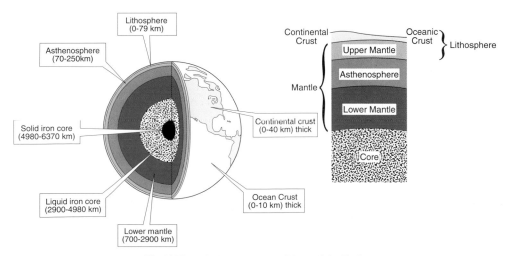

Fig 14 The principal interior divisions of the Earth.

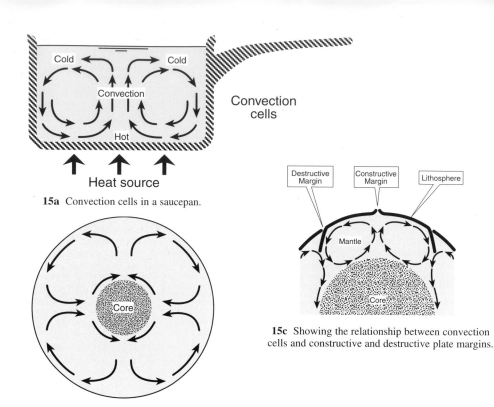

15a Convection cells in a saucepan.

15b Convection cells in the Earth's mantle.

15c Showing the relationship between convection cells and constructive and destructive plate margins.

On a large scale therefore it seems that the movement of plates is part of an overall convective system, with upward movement of melted mantle occurring under constructive plate margins followed by a sideways or lateral movement, and then a descent of the plate into the mantle over the downward limb of the convection cell. The process is summarised in Fig. 15d. Earthquakes and volcanic eruptions occur at both constructive and destructive margins, but while the activity at constructive plate margins tends to be relatively subdued in character, the volcanic eruptions and earthquakes associated with destructive plate margins, such as the Andes and Indonesia, are among the most violent and destructive in nature.

Superimposed on this large-scale circulation is a smaller-scale system of upcurrents which are often referred to as MANTLE PLUMES. These mantle plumes or "hot-spots" account for the volcanic activity on some oceanic islands remote from plate margins such as Hawaii, and they play an important role in starting the process of continental break-up during plate tectonic activity. As mantle plumes reach the surface they cause the crust to dome upwards and the stress created by this doming is relieved by breaking or fracturing of the crust. In the case of the volcanic activity in north-east Ireland and western Scotland it seems that a new set of FRACTURES aligned northwest-southeast were imposed on an older set of north-south fractures, and that movement along

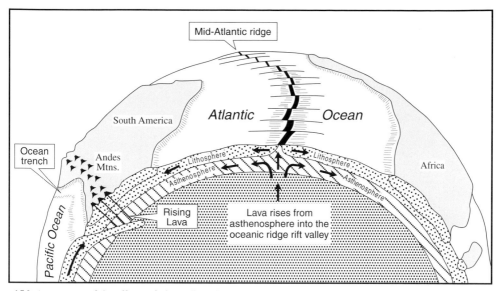

15d A summary of the effects of plates diverging at constructive plate margins and converging at destructive plate margins.

these lines of weakness in the crust allowed the melted mantle material known as MAGMA to make its way upwards to the surface. Splitting or rifting of the continental landmass started with basaltic flood lavas erupting on the continental edges in eastern Greenland and in Ireland and Scotland. Spreading of the North Atlantic also occurred in the Labrador area separating Greenland from north-east Canada with similar volcanic rocks occurring on the west coast of Greenland. Fig. 13b shows this rifting and the main areas of volcanic rocks associated with these events. The spreading process is continuing with a hot spot or mantle plume on the mid-Atlantic Ridge at Iceland, thus accounting for the frequent volcanic activity in that region.

3

Causeway Coast Volcanism

3. Causeway Coast Volcanism

3.1 The First Phase.

By around 65 million years ago at the beginning of the Tertiary period Ireland had emerged from the shallow Cretaceous sea and was a rolling landscape of weathered limestone with pockets of woodlands. The catastrophic events in the southern hemisphere which had brought about the break up of the super-continent Gondwanaland were now about to spread increasingly northwards to break up the northern super-continent Laurasia. The opening and widening of the north Atlantic created a rift between Europe and the east coast of Greenland, and between north America and the west coast of Greenland, with active volcanoes associated with the continental splitting (Fig. 13b). It was this splitting between Europe and Greenland that formed the Antrim basalts. As the crust is stretched prior to rifting it becomes thinner and so mantle rises to fill the space. This mantle material is now at a level in the crust at which it can melt and thus large volumes of magma are generated.

The eruptions in Antrim were part of a series of lava fields which are found from the north-eastern corner of Ireland to the Inner Hebrides of Scotland and as far north as the Faroe Islands and Jan Mayen Island. They represent the first phase of the igneous events associated with this period of crustal rifting. Following the FISSURE ERUPTION of these mainly basaltic lavas, igneous processes tended to be concentrated around more localised CENTRAL VOLCANOES and the great central volcanic complexes of Mull and Skye in the Hebrides with their wide variety of INTRUSIVE and EXTRUSIVE IGNEOUS ROCKS ranging in composition from basaltic to granitic were formed during this later phase. The intrusive complexes of the Mournes and Slieve Gullion also date from this period following the formation of the basalt plateau further north in Antrim.

The onset of volcanic activity in northeast Ireland was marked by explosive eruptions that deposited thick layers of ASH in many places along the north coast. This type of activity can be seen today in Iceland while ancient ash layers can be seen in places like Carrickarade or Kinbane Head (Fig. 16). The explosive phase was then rapidly followed by the widespread eruption of basalt lavas which flowed from long cracks or fissures which opened in the crust in a northwest-southeast direction as a result of the immense forces in the underlying mantle which

Fig 16 Volcanic ash at Carrickarade.

were pulling the crust apart. These cracks or fissures were often filled by magma which when cooled formed DYKES or wall-like intrusions. There are many thousands of them in Ireland, including many outside the area of the basalt lavas and they are most commonly aligned in a northwest-southeast direction which was the predominant direction of fissures in this area during eruption. This type of fissure eruption is often seen today in Iceland (Fig. 17) and the Antrim landscape at that time would have been very similar to central Iceland at the present time, i.e. a barren lava desert with some of the flows shining like black glass and looking like flows of congealed treacle, while other flows resembled huge piles of boiler clinker. Over the whole region during and after eruptions there would have been a pall of sulphurous fumes and dust clouds blown out by the many fissures often with still active volcanic cones as shown in Fig. 18.

Fig 18 Post-eruption landscape in northern Iceland.

The remains of many of the largest of these fissures are still to be found in the landscape of Antrim. Probably the best known is Slemish, a prominent long and narrow hill in mid-Antrim which lies northwest - southeast and is the remains of the lower part of one of the largest fissure systems which included a lava lake like those frequently seen in Hawaii.

3.2 The Building of a Lava Plateau.

These eruptions gradually buried the Chalk landscape under layers of basalt which overlapped each other and built up the first stage of an extensive lava plateau

Fig 17 Active fissure eruption in Iceland.

23

comprising the LOWER BASALTS. Pauses in the eruption sequence allowed the tops of flows to become weathered, and because of the high iron content of basalt lavas, this weathering is often a reddish or purple colour, similar to the process of rusting in iron metal. These red or purple layers allow the separate flows in a sequence to be identified and this shows that flows in the Lower Basalts are generally not more than 10m thick. The tops of the flows often have a thin fine grained layer of weathered dust on the top surface. This dust was probably produced by lava fountains that existed during the eruption phase, the fine lava droplets freeze to a volcanic glass in the air and fall as a glassy dust which then weathers very quickly to give the characteristic thin bright red layer on the lava top.

There are other features which mark the top and base of lava flows, notably the presence of what were formerly gas bubbles. All lavas contain a certain amount of dissolved gases such as water vapour, carbon dioxide and sulphur dioxide. As the lava comes to the surface the reduction in pressure and temperature releases these gases from solution and they form bubbles within the lava. While the interior of the lava is still fluid these bubbles will tend to rise towards the upper surface where more rapid cooling of the top of the flow in contact with the air has caused a crust to form. This hardened crust will trap the rising bubbles, or VESICLES as they are called, so that the upper zone of the lava flow has a concentration of gas bubbles and may be described as VESICULAR. Reaction between groundwater or vegetation and the hot magma may also cause vesicles to form at the base of the flow. During the process of solidification of the lavas many of these empty bubbles are filled with a group of minerals called ZEOLITES which crystallise from hot groundwater circulating through the cooling lava pile. These vesicles filled with zeolite minerals, which are often white in colour, are referred to as AMYGDALES and so the base and top of the basalt flow is often described as AMYGDALOIDAL. The Lower Basalt flows near Windy Gap at the Giant's Causeway show typical lava junctions (Fig.47, section 5.4).

It is not known how long this first phase of volcanic activity lasted but it must have been a period of at least 100,000 years with gaps between over-lapping flows of many hundreds or perhaps thousands of years to allow weathering to take place on the cooled surface of the most recent flows. In addition, trees and other vegetation had time to colonise this seemingly hostile world, because in places pockets of charred vegetation are found between flows. What is known is that as the frequency of eruption began to slow, there were longer and longer periods between eruptions producing thicker and more frequent weathered horizons, and eventually over the whole of Antrim volcanic activity ceased and a long period of dormancy began.

3.3 The End of the First Phase.

This period of dormancy with little or no volcanic activity was in a warm or subtropical climate with abundant rainfall which formed a deep weathered layer on the Lower Basalt surface. This deep weathering under warm conditions produced a thick red "interbasaltic" layer which is similar to soils that are found today in many tropical areas such as central Africa and is known as LATERITE. The alteration of the basalt lavas by the processes of weathering reduces the silica content and leaves the laterite enriched in iron and aluminium, such that the inter-basaltic layer in Antrim has been mined for both iron ore and aluminium ore (BAUXITE) during the recent past. The Interbasaltic surface supported vegetation and there are occasional beds of LIGNITE present, formed from accumulations of plant debris. Lignite is a form of coal and some of these deposits were thick enough to be worked commercially, such as those from the Craignahulliar Quarry, near Portrush. The laterite layer is often 10-15m thick and is found extensively over north and mid-Antrim indicating that the period of quiescence or dormancy was widespread over the whole volcanic province of northeast Ireland, although there were interruptions in places. As with the preceding volcanic phase it is not certain how long this Interbasaltic period lasted, but it was certainly tens of thousands of years at least, long enough for the first two or three flows of the Lower Basalt surface to be rotted completely, and also long enough for deep river valleys to be cut on the Lower Basalt landscape. These river valleys were to play a major role in the next phase of the volcanic history of Antrim, the eruption of the Causeway basalts.

3.4 The Eruption of the Causeway Basalts.

During this prolonged period of volcanic inactivity the landscape of the Lower Basalts was radically changed from the black barren ash-covered terrain of the eruptive stage to one with lakes and rivers and vegetation which included conifers such as cedar, spruce and pine and deciduous trees such as hazel and alder. The existence of the lakes and rivers are known from the often steep-sided valleys seen on the Lower Basalt surface and from

deposits of fine-grained SILTSTONES and mudstones occasionally left preserved on the laterite. Pockets of lignite have yielded plant remains in the form of leaf marks and pollen grains which have allowed identification of the species which grew on the Interbasaltic surface.

There had been intermittent small scale volcanic eruptions since the end of the main eruptive phase which produced the Lower Basalts. These occurred mainly in mid-Antrim, producing for example the volcanic complex at Tardree consisting of pale-coloured RHYOLITE, and the volcanic glass found at Sandy Braes, both near Ballymena. These rocks are quite different in composition from the Lower Basalts, but are chemically derived from them and can be considered as the last waning stages of the Lower Basalt eruptions.

In north Antrim however there began a much more extensive volcanic event which produced a set of basalt lava flows very different in form and appearance from the Lower Basalts which preceded them. The lava flows of the Causeway basalts are distinctive from the Lower Basalts by being generally much thicker, frequently they are up to 30m thick, while the flow which contains the Causeway itself is around 100m thick. They are also famous for their COLUMNAR JOINTING and although all basalts form columns to some extent, the other lavas in Antrim do not show the remarkable Iregularity of those at the Causeway (Fig. 19).

The Causeway basalts are restricted to a relatively small area of north Antrim, mostly north of a major FAULT line, the Tow Valley Fault, which runs south-west across north Antrim from the coast at Ballycastle. This fault is an important zone of crustal weakness that can be traced east-

Fig 19 Columnar Causeway basalts overlying red laterites of the Interbasaltic Formation above Lower Basalts.

SILLS and also of course the lavas which extrude on to the surface. Movement along the Tow Valley Fault and the Foyle Fault, which parallels it to the north-west, is thought to have initiated the production of the Causeway basalt magma.

When this next phase of lava eruption began after the prolonged interlude of the Interbasaltic period, the landscape had much steeper and more extensive valleys

wards to the Scottish Highlands and westwards to the coast of County Mayo. There is evidence that this fault zone has been intermittently active for the last 500 million years at least and it is likely that the eruption of the Causeway lavas was linked to movement on the fault during Tertiary times. The effect of crustal rifting such as was taking place during the eruption of the Antrim lavas is to place the crust under tension, or to stretch and thin it. This causes a gradual reduction in pressure in the mantle underlying this thinner crust and makes it more susceptible to melting. If the crustal tension is relieved by the sudden movement along faults such as the Tow Valley Fault then the fall in pressure thus brought about can lead to large-scale melting of the mantle beneath, producing great volumes of basaltic magma. As this liquid makes its way to the surface it forms a reservoir of magma which feeds the various intrusive features such as dykes and

and hollows than the Cretaceous Chalk landscape which had been buried by the eruption of the Lower Basalts. In addition, although the result of fissure eruptions like the Lower Basalts, the scale of the fissures and the rate of flow of magma from them seems to have been much larger for the Causeway Basalts. This greater volume of magma and the deep valleys on the Interbasaltic surface meant that the fairly fluid lava formed deep lava ponds as the irregular landscape was levelled by the first great outpouring of lava. The Giant's Causeway formed in such a valley with the first flow here attaining a thickness of almost 100m (Fig. 20).

3.5 The Causeway.

The most striking feature of the Causeway lavas and the one for which they are world famous, is their tendency to form columns, frequently of remarkable regularity and complexity.

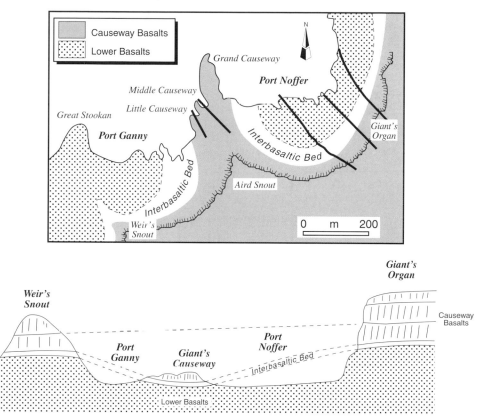

Fig 20 Geological map and cross-section of the Giant's Causeway locality (after Wilson and Manning, 1978).

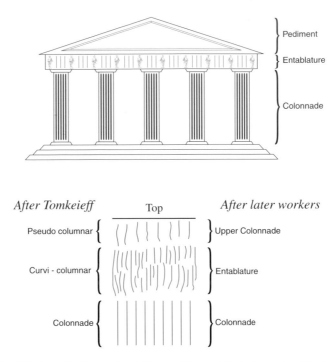

Fig 21 Sub-divisions of columnar basalt flows and their origin in classical architecture.

The paintings by Susanna Drury were the first attempt to accurately represent the detail of the Causeway jointing, while in 1940 a Russian-born geologist S.I. Tomkeieff published a scientific paper which suggested descriptive terms for the Causeway jointing. These have since passed into the geological literature and are used to describe columnar basalts all over the world (Fig. 21).

Tomkeieff recognised a three-fold or tri-partite division in the columns of a typical Causeway lava. Using terms borrowed from classical architecture he called the lower division, consisting of regular vertical columns, the COLONNADE, with an upper less regular zone called the ENTABLATURE. This upper zone he divided into a curvi-columnar zone of narrow, often curved columns which passed up into a zone of wide and less obvious columns called the pseudo-columnar zone. The colonnade is considered the equivalent of the roof supporting columns in a classical Greek temple, while the entablature is that part of the roof structure which commonly has a frieze with elaborate carvings.

The proportions of the various subdivisions within flows can vary, but no one zone can be considered in isolation, they are all part of the internal structure of the flow. Later research workers have referred to the pseudo-columnar zone of Tomkieff's nomenclature as the upper colonnade as in fact it forms in the same way as the lower colonnade.

The Giant's Causeway itself is the colonnade of the first flow of the Causeway basalts (Fig. 22), and at the locality known as the Giant's Organ on the east side of Port Noffer, this same colonnade can be seen passing up into the entablature zone of the first flow (Fig. 23 and 52).

28

Fig 22 The colonnade of the first flow of the Causeway basalts at the Giant's Causeway.

3.6 Formation of Columnar Jointing.

All igneous rocks are jointed to a greater or lesser degree, the joints are either tectonic, that is caused by the deformation of the earth's crust, or are caused by shrinkage or contraction of the semi-solid interior of the lava flow during cooling. Many igneous rocks will show both of these types of jointing but in the Causeway lavas the overall jointing pattern is predominantly due to shrinkage after cooling. The polygonal pattern produced in mud in a dried-up pond is caused by shrinkage due to drying of the mud and is a very similar phenomenon, although it occurs on a much smaller scale and only effects the surface layers of mud.

Lava flows like those at the Causeway are erupted at temperatures of around 1100 degrees C and lose heat very rapidly to their surroundings. Near horizontal features such as lavas lose most heat through their top and bottom surfaces. The internal stresses set up by thermal contraction produce sets of roughly parallel joints or frac-tures at right angles to the cooling surfaces and these joints, moving inwards towards the centre of the flow as it solidifies, take the form of a "cooling front" (Fig. 24a). Cracks are started at many points on the surface of this cooling front and three-pronged cracks at angles of about 120 degrees occur at each of these points. As these cracks propagate they intersect to form irregular polygons of 3, 4, 5, 6 or 7 sides (Fig. 24b). With continued cooling of

Fig 23 The locality known as the Giant's Organ showing the colonnade of the first flow passing up into the entablature.

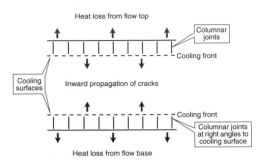

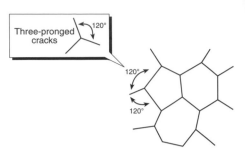

Fig 24a Horizontal cooling surfaces in a lava flow lead to the formation of vertical parallel columnar joints.

Fig 24b Showing the effect of intersecting cooling cracks formed at 120° to each other forming irregular polygons.

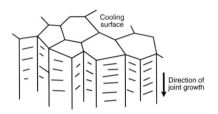

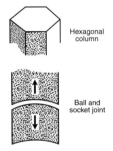

Fig 24c Formation of polygonal columns as solidification proceeds and the cooling cracks move inwards.

Fig 24d Cooling contraction along the length of columns forms the "ball and socket" joints which divide the vertical columns horizontally.

the lava flow these polygonal cracks, which started at the top and base of the flow, will move inwards as solidification proceeds to form 3-dimensional polygonal columns, roughly at right angles to the horizontal cooling surfaces (Fig. 24c). The columns also shrink along their length and this contraction produces the convex and concave "ball and socket" joints which divide the vertical columns horizontally (Fig. 24d).

In contrast, if the magma bodies are near vertical, such as in dykes, which are the lower parts of the fissures which fed the lava flows, then most heat is lost through the side walls rather than through the top and bottom surfaces. This means that in these cases the cooling surfaces are close to vertical and so cooling joints in dykes are typically near horizontal (Fig. 24e).

Simple columnar jointing is common in many Antrim lavas and intrusions. The special attraction of the columnar jointing in the lavas of the Giant's Causeway area lies in the spectacular and multi-tiered form it commonly takes and the fact that the jointing can be readily examined in the horizontal and vertical planes. The formation of this type of complex layering appears to be due to the surface of a thick body of cooling lava being flooded or inundated by large volumes of water. This water passes down into the hot, still crystallising flow and disrupts and modifies the cooling

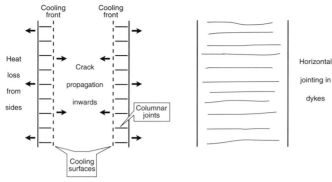

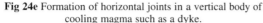

Fig 24e Formation of horizontal joints in a vertical body of cooling magma such as a dyke.

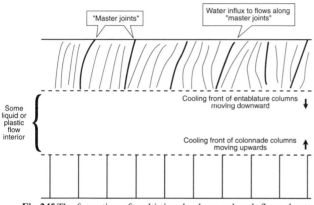

Fig 24f The formation of multi-tiered columnar basalt flows due to the flooding of the flow surface during cooling.

The typical sequence from top to bottom of a Causeway flow, i.e.

i   upper colonnade or pseudo-columnar zone

ii curvi-columnar zone

iii colonnade

can be explained by an initial period of "dry" cooling to form the lower colonnade with columns growing up from the base of the flow and the upper colonnade with columns growing down from the upper flow surface. This initial dry cooling is followed by a period of inundation of surface water to give "wet" cooling and chilling in the upper part of the flow such that the upper colonnade type of columns gives way to the narrower and often curved columns, with a more rapid growth rate, which form the entablature. Water displaced on to the surface of a still hot lava flow will migrate downwards through early-formed joints on the surface of the flow. Called "master joints" these can be seen clearly as wider and more persistent cracks in many places through the flow, cutting the entablature with a spacing of 3-5m.

Observation of the surface of ponded magma bodies such as Hawaiian lava lakes has shown that polygonal contraction cracks start to form within minutes of the lava beginning to cool. These cracks continue to move downwards for the entire cooling period of the lava unless disrupted

processes within the flow. The effect of this disruption is to speed up the cooling in the upper part of the flow to produce the curvi-columnar zone of the entablature formed by narrow columns growing rapidly downwards, with the more regular columns of the colonnade growing upwards from the base of the flow at a slower rate (Fig. 24f).

In effect the two cooling fronts within the flow produce contrasting styles of columns because of the addition of large quantities of water to the upper part. The very sharp junction between the two column types as seen at the Giant's Organ (Fig. 23 and 52) is the point within the flow where the more rapid entablature front moving down met the slower colonnade front moving up.

by further influxes of magma into the lake. The size of the polygons formed is 5-10m across, i.e. similar to the scale of the master joints observed in the entablatures of the Causeway flows.

The variation in the relative proportions of colonnade and entablature is due to the amount of water entering from the surface of the flow during the cooling phase, and

Fig 25 Example of "flow-foot breccia" at Port na Spaniagh caused by hot lava advancing into water.

the occasional absence of curved columns from a flow (e.g. flow 2, west of Port na Spaniagh), probably indicates that not all parts of the flow surface were flooded, with some areas unaffected. At this particular locality therefore only columns of the upper and lower colonnade types formed.

Clearly then there is a need for substantial quantities of water to be present during this period for this theory of column formation to be correct. There are a number of volcanological features within the Causeway lavas that indicate the presence of water, either as rivers or lakes, during the eruption and cooling of the lavas. The flow forming the Causeway was recognised by Tomkeieff as having ponded in a wide river valley cut into the Interbasaltic surface. Major sheet floods of lava such as those which produced the Causeway flows would have had the effect of damming such river valleys and displacing the water elsewhere, either as widespread shallow lakes or into new river courses. There are numerous clear examples that this happened during the forma-

tion of the much younger Columbia River basalt plateau in the western USA, where the Columbia and Snake Rivers were frequently diverted by sheet floods of lava similar to those around the Causeway. These former valleys are now marked by extensive river-derived sediments between lava flows, or by dry canyons which no longer contain rivers. The fine grained sandstones and SHALES with plant remains, found between flows of the Causeway basalts in Craignahulliar Quarry near Portrush are similar to deposits formed in rivers or shallow lakes and show that surface water was abundant in the environment of eruption of the Causeway lavas.

When a lava flow moves from land into water, whether lake, sea or river, the freezing of the hot lava against the cold water produces fragments of volcanic glass known as HYALOCLASTITE. These glass fragments are often built up into delta-like forms in front of the advancing lava flow and are sometimes referred to as FLOW-FOOT BRECCIAS.

Occasionally larger masses of lava are preserved within these breccias in the form of rounded globules or PILLOWS. Such formations are found at the base of several of the Causeway flows, e.g. the base of flow 1 east of Port na Spaniagh (Fig. 25). Note that at the time of writing the Port na Spaniagh area is inaccessible; see comments at the start of section 5.4. The breccia consists of pillow-like basalt masses surrounded by formerly glassy fragments which have now been altered by weathering to a yellow/orange coloured clay material known as PALAGONITE. Similar features can be examined at locality 3b, section 5.3.

It is significant that columnar flows elsewhere, at Fingal's Cave on Staffa for example, are also found with pillow lavas, hyaloclastite breccias and lake and river sediments.

3.7 The Final Phase.

After the eruption and cooling of the Causeway lavas there was another period of dormancy over the whole of north-east Ireland and this is shown by the existence of a second layer of laterite that occurs on top of the Causeway basalts at places like Croaghmore Hill, south of Whitepark Bay, or at Ballylagan, near Portrush. This volcanic quiescence was interrupted by the eruption of what was to be the final stage in the building of the Antrim lavas, the UPPER BASALT formation. This phase represented a return to lava types and volcanic conditions similar to those of the Lower Basalts, with thinner flows than those of the Causeway lavas, and a lack of · well formed columns. The full lava stratigraphy of north Antrim is shown in Fig. 26. In the Causeway coast area the Upper Basalts are restricted mostly to the southern part of the area with a faulted block on the coast near Portbraddan (see geological map, Fig. 4). There is no way of knowing the full extent of the Upper Basalt lavas erupted, but it is likely that in the last 60 million years or so a considerable thickness has been removed from the top by erosion.

3.8 Summary of the Sequence of Events.

Figure 27 shows in simplified form the sequence of events during the formation of the various components of the Antrim Lava Group.

a) By the end of Cretaceous times and the beginning of Tertiary times a low rolling landscape had developed on the Chalk surface with sparse drainage and pockets of vegetation.

b) With the beginning of continental rifting associated with the opening and widening of the North Atlantic Ocean, fissures developed on the Chalk land

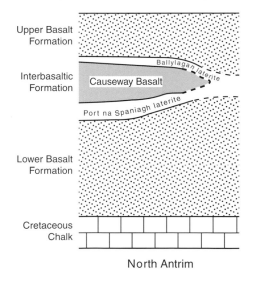

Upper Basalt Formation

Interbasaltic Formation — Causeway Basalt — Ballylagan laterite

Port na Spaniagh laterite

Lower Basalt Formation

Cretaceous Chalk

North Antrim

Fig 26 The stratigraphy of the Antrim Lava Group in north Antrim (after Lyle and Preston, 1993).

a) the late Cretaceous - early Tertiary landscape some 65 million years ago

Pockets of vegetation

Rivers & Lakes

Cretaceous Chalk

b) continental rifting begins

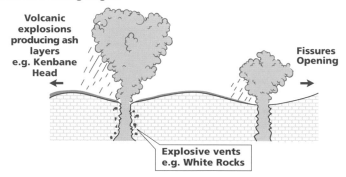

Volcanic explosions producing ash layers e.g. Kenbane Head

Fissures Opening

Explosive vents e.g. White Rocks

c) fissure eruption of the first, Lower Basalts - the building of a lava plateau

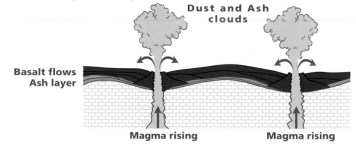

Dust and Ash clouds

Basalt flows
Ash layer

Magma rising   Magma rising

d) end of the Lower Basalt development - plant colonisation and weathering of the basalts to form the lower red laterites and river valleys

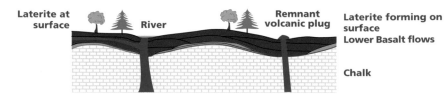

Laterite at surface   River   Remnant volcanic plug

Laterite forming on surface
Lower Basalt flows

Chalk

e) fissure eruption of the
Causeway Basalts.
Formation of shallow lakes
by disruption of rivers
important in the formation
of columnar basalts

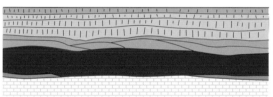

← Fissures opening **→**

Giant's Causeway

Entablature — Water on surface / Causeway Lavas
Colonnade — Inter - Basaltic / Lower Basalts / Ash layers / Chalk

Magma rising

f) end of the Causeway Basalt development -
weathering of the basalts to form the upper
red laterites

Upper Interbasaltic Laterite
Causeway Lavas
Lower Interbasaltic Laterite
Lower Basalts
Chalk

g) fissure eruption of the
Upper Basalts - the final
phase of volcanic activity

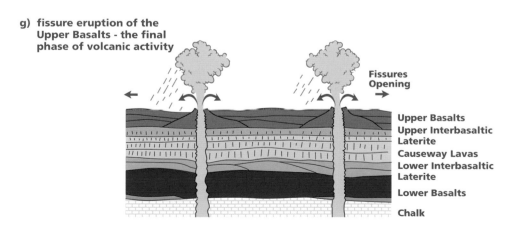

Fissures
Opening
← **→**

Upper Basalts
Upper Interbasaltic
Laterite
Causeway Lavas
Lower Interbasaltic
Laterite
Lower Basalts
Chalk

Fig 27a–g The sequence of events during the formation of the Antrim Lava Group in north Antrim.

35

surface and layers of volcanic ash such as those at Kinbane Head and Carrickarade were erupted.

c) Following this initial explosive stage the Lower Basalts were erupted from long fissures which opened up across the whole of north-east Ireland. These eruptions built up the lava plateau with thin volcanic ash and dust layers between flows.

d) After the eruption of the Lower Basalts a prolonged period of dormancy began with little or no volcanic activity in the area. During this stage the climate was wet and relatively warm and the basaltic land surface was deeply weathered to produce bright red lateritic soils. There were also major rivers flowing across the land forming deep and wide river valleys, with a range of tree species growing including alder, cedar and pine.

e) Renewed volcanic activity produced the Causeway lavas in north Antrim. These flows were thicker than the earlier Lower Basalts and in places such as the Giant's Causeway they completely filled the wide river valleys that had been carved on the Lower Basalt surface during Interbasaltic times. This displaced the river water which flowed over the hot surfaces of the Causeway flows and formed the colonnade and entablature columnar jointing which is typical of their appearance.

f) After the eruption of the Causeway lavas there was another period of dormancy, probably shorter than that following the Lower Basalt eruptions, and the upper Interbasaltic laterite was formed.

g) Once again volcanic activity renewed with the development of fissures on land in Antrim and the eruption of the Upper Basalts. These were formed in a similar way to the Lower Basalts and marked the end of volcanic activity in north-east Ireland as the mantle plume or hot spot producing the magma shifted its main centre of activity elsewhere in the north Atlantic area.

4
Quaternary

4. The Quaternary

4.1 Quaternary

The end of igneous activity was followed in north-east Ireland by a period of earth movements with widespread warping of the crust to give a general basin or SYNCLINE form to the lava plateau. This basin was centred around Lough Neagh and erosion of the surrounding higher ground produced large quantities of sediments known as the Lough Neagh Clays which are OLIGOCENE in age (Fig. 5), the youngest of the solid geology rocks of north-east Ireland. These beds are up to 300m in thickness and consist of mostly sands and clays. In a number of areas including Crumlin and Ballymoney, these beds contain substantial deposits of lignite, more than 50m in thickness around Crumlin for example.

The general absence of PLIOCENE (upper Tertiary) and Lower Pleistocene beds in the area shows that phases of erosion continued into the Quaternary. The most significant events during the Pleistocene were the development of large mid-latitude ice sheets. The effects of these were global and were especially significant over much of the northern hemisphere when erosion and deposition by the ice moulded the landscape of Ireland in ways which are still obvious today. Most of our landform patterns are related directly to the last phase of glaciation that took place between 25000 - 17000 years ago.

The last 2 million years is known as the Quaternary and during this period the climate fluctuated between warm (inter-glacial) and cold (glacial) stages. Shorter climatic spikes of a sub-Arctic or BOREAL aspect are known as INTER-STADI-ALS. During the glacial stages ice sheets waxed and waned over wide areas and during the warm inter-glacial stages temperatures were as high or indeed slightly higher than at present. Large-scale ice sheets in mid-latitudes such as the British Isles only developed during those parts of the cold stages when the climatic conditions where able to supply sufficient moisture for ice sheet growth. For the remainder of the time Arctic-type deserts such as those currently present in northern Canada or Siberia were the rule. At least 20 cold episodes have been recognised in Europe largely from fossil evidence from plants, insects and animals recovered from inter-glacial sediments and there is no indication that these climatic fluctuations have finished. It is likely that we are currently living through what is the most recent inter-glacial period and, in the normal course of geological events, could expect a return to glaciation, global warming due to human activity permitting of course.

Since it is estimated that some of the inter-glacial periods lasted as long as 90,000 years, and the last major ice sheet left north-east Ireland less than 20,000 years ago, we could be in the early stages of an inter-glacial period. Just as climatic oscillations have been recognised in the glacial stages, so it has been possible to recognise climatic variability in the inter-glacial stages. The cyclical change from Arctic to boreal to temperate conditions, returning to boreal and Arctic has been recorded in inter-glacial lake sediments and other deposits. Current evidence suggests that climatic changes of these types can occur very rapidly indeed.

The detailed interpretation of the glacial history of north-east Ireland is difficult and at times dependent on variable interpretations of the evidence. It involves ice sheet movements from lowland Ireland

Fig 28 Basal glacial clays with overlying shallow-water marine and raised beach deposits.

glacial sediments are termed ground moraines and lateral moraines form along the edges of valley glaciers in upland zones.

Following the advance of the Scottish ice, the next major event had its centre on Lough Neagh, with ice moving south-east and northwards into the Malin Sea off County Donegal. This advance resulted in a major phase of DRUMLIN formation and these elongated hills are very characteristic of lowland glaciated areas in the north of Ireland. Drumlin long axes tend to parallel the former directions of ice movement. Around 16,000 years ago a re-advance or surge of Scottish ice occurred some way into north-east Ireland and formed a moraine between Limavady and Ballykelly. The final event of the Midlandian Cold Stage, approximately 14,000 years ago, was the formation of late-glacial raised beaches underlain by shallow-water marine deposits, such as those exposed around Portballintrae harbour. (Fig. 28).

and western Scotland together with locally generated ice sheets and small valley glaciers. The north-east of Ireland has probably been affected by a number of glacial events though the lack of field evidence for older episodes may be the result of erosion by more recent ice sheet events. The last cold stage, the DEVENSIAN (MIDLANDIAN), lasted from approximately 100,000 to 10,000 years ago. The events which have left the greatest impact on the landscape of north-east Ireland date from the last 25,000 years, in the late Midlandian. Multiple ice sheet oscillations are documented.

Initially there was an advance of Irish ice northwards from central Ulster followed by an advance from the northeast by Scottish ice to form the glacial deposits at Armoy, between Ballycastle and Ballymoney. These elongated ridges, which mark the former ice margin, are termed MORAINES and are largely the result of bulldozing or thrusting by the ice of sediments which had been deposited in front of the ice sheet. Flat areas of sub-

The landforms or GEOMORPHOLOGY resulting from the last major phase of glaciation are the single most dominant feature of the present day landscape. The extent of the influence of ice from Scotland is shown by the occurrence of blocks of a fine-grained granitic rock transported from the prominent island of Ailsa Craig in the Forth of Clyde. These non-local rocks are known as ERRATICS and the Ailsa Craig material has a very distinctive blue-speckled appearance. Their presence in the glacial sediments or TILL along the east coast and as far inland as Tyrone and Armagh is indicative of the direction in which the ice was travelling. One of the most widespread of glacial deposits is BOULDER CLAY which formed below the ice as ground moraine and is a stiff clay, often containing boulders of all sizes, and formed by the plucking and abrasive action at the base of moving ice. The last phase of ice movement from the Lough Neagh Basin northwards onto the continental shelf sculpted many of the drumlins observed in north-eastern Ireland.

During periods of ice melting, deposits of sand, silt and clay were transported by rivers which flowed in tunnels in the ice. These discharged material onto the area in front of the ice. In ponded water these outpourings often formed delta-like features such as those seen in the Ballyvoy area of Ballycastle, immediately to the east of the Causeway coast AONB. Often deposits consisting of jumbled or chaotically bedded sand and gravel, which occur in the form of ridges called ESKERS, mark the course of these sub-glacial channels. They are a valuable source of sand and gravel for the construction industry over the whole of Northern Ireland. However the exploitation of this non-renewable resource is currently a matter of serious conservation debate.

During the melting phase of the last ice sheet the catastrophic release of large volumes of water resulted in the formation of temporary glacial drainage channels and impounded glacial lakes. The largest example within the AONB occurs at Lisnagunogue near the Causeway Visitor centre. This valley is about 4 km long and 300m wide and currently carries only a MISFIT STREAM but must have carried a considerable volume of water during its formation as a glacial meltwater channel.

As the ice-sheets disappeared at the end of the Midlandian Cold Stage, but prior to the current inter-glacial period, the area experienced a tundra landscape and climate. This phase was climatically and ecologically similar to that presently found in the Canadian or Russian Arctic, with a characteristic vegetation dominated by grass and sedge, dwarf birch and dwarf willow trees. The climate warmed sufficiently to allow grassland to develop with the widespread occurrence of the Giant Irish Deer whose enormous antlers (up to 3m wide) have been found all over Ireland. A short-lived reversion to cold conditions brought about the disappearance of the grasslands, and with the grasses went the Giant Deer, unable to survive as temperatures fell and the vegetation became increasingly sparse. Finally about 10,000 years ago a general climatic warming heralded the beginning of the current warm stage (Holocene) which has been relatively stable climatically.

4.2 After the Ice Age.

The landscape modifications of greatest geological significance to have affected the Causeway Coast area since the disappearance of the great ice sheets are the coastal erosion and deposition features associated with post-glacial sea level changes and storm events, the development of areas of blown sand and the formation of peat.

The fluctuations in sea-level following the retreat of the ice were caused by the variable interactions of two processes; the readjustment of the earth's crust after

Fig 29 Modern beach at White Rocks, Portrush.

2-3m above present day possibly around 6,000 years ago. Much of the present coastal scenery derives from the period after this sea-level peak. Sediment was transferred onshore, forming for example the beach ridges and dunes at Portrush, Runkerry and White Park Bay, and the raised shoreline fronting the cliffed coast around the Giant's Causeway and further east at Ballintoy Harbour.

The extensive sandy beaches at Portrush, the Bushfoot Strand between Portballintrae and Runkerry and White Park Bay, are backed by areas of blown sand consisting mostly of quartz grains from the re-working of glacially-derived sediments. (Fig. 29).

The essentially progressive improvement of climate since 10,000 years ago, enabled forest trees to return gradually to north-western Europe with widespread tree coverage developing over Ireland. Woodland diversity in Ireland probably reached a peak around 7,000 years ago with a wide range of plant species and a varied fauna associated with them including deer, boar, wolves and fox. The climate at this stage was probably slightly warmer than at present. However a climatic deterioration with a change to a more humid and wetter climate meant that conditions were suitable for the development of peat bogs. A rise in the water table meant that ground that had previously supported trees now became waterlogged and

removal of the ice load and an increase in the volume of water in the oceans due to melting of the large continental ice sheets. The vast weight of the ice pressing down on the crust over thousands of years had depressed it. Subsequent removal of the ice load during deglaciation allowed the crust to rise gradually. However, due to the relatively short period over which the ice caps melted, the water level rose faster than the crust. This meant that, in the short term, late-glacial sea-levels were much higher than at present, possibly up to 100m above current levels). These higher sea-levels have given rise to coastal features such as raised cliffs, STACKS, arches, caves and beaches at various levels above the present High Water Mark (see Fig.59, section 5.5.1). Commonly, RAISED BEACHES in north Antrim reach to +21m, for example on Rathlin Island.

As the land rose, the water level fell steadily to a low point of perhaps 100m below present levels about 11,000 years ago. About 10,000 years ago sea levels began to rise rapidly, peaking about

a layer of peat material buried the former forest floor. This environment was wet and slightly acid which therefore slowed down the rate of vegetation decay and allowed the accumulation and of raised bogs. This marked shift from woodland to bog was partly influenced by forest clearance undertaken by people to permit agricultural development, the beginning of mans modification of the Irish landscape. This new force is now dominant in shaping the world around us – but only in human timescales. Geological processes continue un-abated with plate tectonics indicating the magnitude of natural phenomena which humans are powerless to control.

These final stages in the geological development of the landscape are now being modified by human activity. Changes which occurred over the past 9,000 years have fashioned the landscape with which we are familiar. The balance of the landscape is complex with the geological framework providing the basis for the current biological diversity and human occupation.

5

Excursion Guide To The Geology Of The Causeway Coast

5. Excursion Guide To The Geology of the Causeway Coast.

The general location of all sites described in this section are given on Fig. 1 while their grid reference, together with summary details and book section and page number are given in Appendix 3. All localities can be found on sheets 4 (Coleraine) and 5 (Ballycastle) of the Ordnance Survey of Northern Ireland 1:50,000 scale maps.

5.1 Portrush.

Locality 1. The Portrush Sill.

All sections mentioned are accessible from the seafront at Landsdowne Cresent, Portrush, where car parking, toilets and other facilities are available. Visitors can walk over Ramore Head by way of a series of footpaths.

The largest outcrop of intrusive rock in the AONB is the Portrush Sill (see geological map, Fig.30) which forms the prominent headland of Ramore Head and underlies most of the town of Portrush to the south (Fig. 31). Due to the historical significance of this locality part of the foreshore has been declared a National Nature Reserve by the DOE(NI). The sill extends offshore and forms an arcuate chain of islands, the Skerries, for about 4 kms to the north-east. It is formed of a coarse DOLERITE and intruded into Lias mudstones which have been baked by the heat of the magma into a hard dark, fine-grained brittle rock known as HORNFELS. These baked sediments, historically known as "Portrush Rock", contain abundant fossils, particularly ammonites. The appearance of fossils in what was thought to be a basaltic rock led the Reverend William Richardson, in the late eighteenth century, to use this occurrence in support of the Neptunist case that basalts resulted

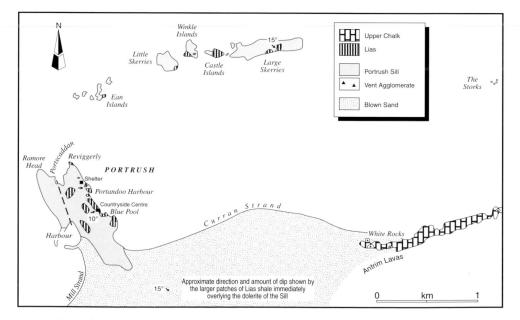

Fig 30 Map showing the geology of the Portrush Sill area (after Wilson and Manning, 1978).

Fig 31 Aerial photograph of the peninsula of Ramore Head which forms the main outcrop of the Portrush Sill.

failed to see the contact between these rocks and the underlying coarser dolerite of the sill. By the early years of the nineteenth century this mistake had been realised and the metamorphic or changed nature of the Lias rocks had been recognised.

The contacts between the dolerite sill and the overlying "roof" of Lias rocks are best seen in a number of places along the foreshore between Reviggerly Point and the Portrush Countryside Centre (see map, Fig. 30). The example shown in Fig. 32

from crystallisation from sea-water (see section 1.2). He had in fact mistaken the baked dark-coloured Lias rocks containing the ammonites for basalt, and

Fig 32 The contact between the dolerite of the Portrush Sill and the fossiliferous Lias shales..

is near the High Water Mark close to the navigational triangle and mid-way between the Shelter and Countryside Centre. The Lias shales have near-horizontal bedding planes which are clearly visible and which tend to weather to a lighter colour than the dark brown weathered dolerite. Many examples of ammonite fossils are exposed on the Lias bedding planes (Fig. 33). Readers are reminded that this is an important conservation site and no attempt should be made to collect specimens by hammering. Loose blocks containing fossil fragments can usually be found with careful searching.

Fig 33 Ammonite fossils on Lias shale bedding planes.

The upper part of the sill is relatively fine-grained where it cooled more rapidly against the Lias rocks. The more rapidly cooling of an igneous rock takes place, the smaller the crystals which form, leading to the formation of a fine-grained "chilled" margin against the rock it is intruded into. Further away from the contact, where cooling is slower, the crystals formed are larger, this difference in grain-size will be noted if the dolerite is examined on Ramore Head. The bottom contact of the sill is not exposed and the maximum thickness seen is estimated at about 45m.

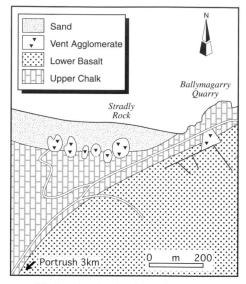

Fig 34 Map showing the geology of the White Rocks area (after Patterson, 1963).

5.2 The White Rocks.

Locality 2. The White Rocks Vents.

The site is best reached by the slip-road, sign posted "White Rocks", which leads off the main Portrush to Bushmills road. There is a car park near the beach. All outcrops can be accessed from the beach.

Around the White Rocks, 3 kms to the east of Portrush, Lower Basalts are exposed on top of the Chalk and there is a complex of small explosive VENTS of both basaltic and chalky AGGLOMERATE. Particularly well exposed examples are to be found to the east of the car park (see map, Fig. 34). This section of the coast is also an ideal location to examine the characteristics of the Cretaceous Chalk or White Limestone of north Antrim. This rock corresponds to the Cretaceous Chalk found in England which forms the famous White Cliffs of Dover, but differs consid-

Fig 35 Elephant Rock at the White Rocks, sea-weathered Cretaceous Chalk on a prominent wave-cut platform.

The precise origin of flint within the Chalk is still controversial but most seems to have formed below the sea bed within the accumulating coccolith ooze.

The Chalk in Antrim can be sub-divided into a number of sedimentary units or members based on variation of lithology, the type of flint nodules present or the presence or absence of hardgrounds. At the White Rocks the Portrush Chalk Member forms the sea cliff above the North Antrim Hardground. The Chalk in this area, because of its resistance to erosion, has formed very good examples of sea caves, stacks and arches, with a well developed wave-cut rock platform at the base of the cliff (Fig. 35).

Starting from the car park at the bottom of the road which leads down to the White Rocks proceed to the beach and walk eastwards in the direction of Dunluce Castle and the Giant's Causeway. At this locality, Stradly Rock, (see map, Fig. 34) which is about 400m across, the Chalk has clearly been shattered and disrupted and then re-cemented by explosive volcanic action which has formed agglomerate-filled vents and deposits of TUFF. Around Stradly Rock there are numerous examples of these vents cutting through the Chalk and into the overlying Lower Basalts. The best example occurs about 150m to the east of the car park and appears to be roughly aligned in a northwest - southeast direction with a near vertical contact with

crably in its physical properties in that it is extremely hard and compacted. This hardening probably took place during burial by the thick pile of basalts that erupted over the Chalk landscape during Tertiary times. The Chalk is formed of the fossil remains of coccoliths (see section 2.3) which accumulated on the ocean floor to form a coccolith OOZE or fine mud. Coccoliths live in today's oceans in the upper levels where sunlight levels permit photosynthesis. At times during the accumulation of this mud there were interruptions in sedimentation caused by factors such as shallowing of the sea. During these periods the soft ooze became hardened and mineralized to form a surface known as HARDGROUND, often with green GLAUCONITE mineral staining which commonly occurs in marine sediments. The presence of glauconite is generally taken to indicate a slow rate of accumulation of SEDIMENTARY particles. A characteristic feature of the chalk is the occurrence of flint in the form of nodules of varying sizes, often occurring in horizontal layers parallel to the Chalk bedding planes (see section 2.3).

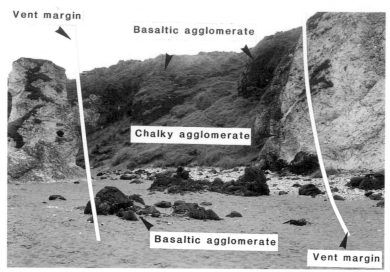

Vent margin

Basaltic agglomerate

Chalky agglomerate

Basaltic agglomerate

Vent margin

Fig 36 Vent with basaltic and chalky agglomerate
at the White Rocks. .

this locality, there is a possibility that elsewhere along the coast apparently similar features are the result of collapse of basaltic material into hollows in the Chalk surface caused by solution of the limestone by weathering. For example features seen at the contact between Chalk and Lower Basalts at the nearby Ballymagarry Quarry (see map, Fig. 34) could be interpreted as collapse structures. This quarry also provides a good example of the weathered Chalk surface with its residual "clay-with-flints" layer which was buried under the Tertiary basalts.

These small-scale explosive vents on the north Antrim coast are concentrated in

the Chalk in the adjoining cliff face (Fig. 36) which shows the steep contact between dark basaltic agglomerate and shattered Chalk. The orientation and the vertical contact with the Chalk on either side suggests that it may represent agglomerate preserved along the line of an eruptive fissure or vent. Fig. 36 shows basaltic and chalky agglomerate at the edge of the vent in the face of the cliff and the shattered nature of the Chalk at the vent margins can be clearly seen. The basaltic agglomerate in the interior of the vent has a coarse blocky character. (Fig. 37).

While there is an undoubted volcanic origin for these features at

Fig 37 Coarse basaltic agglomerate in the interior of the vent..

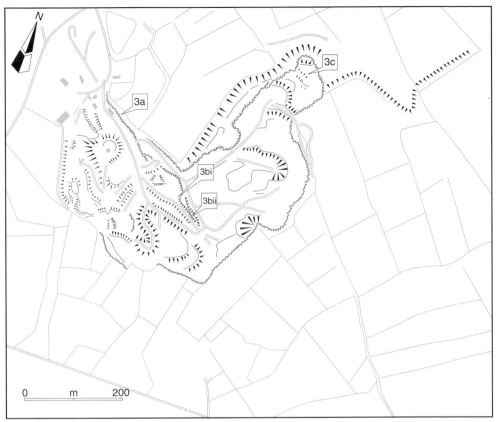

Fig 38 Locality map of Craignahulliar Quarry.

a relatively narrow zone between the White Rocks and Dunluce Castle some 2 kms to the east. The source of the gas to provide the explosive force is likely to be a magma body rising to the surface from depth where the volume of gas was possibly augmented by steam produced by interaction between the hot magma and water contained in the sediments it was passing through. It may be significant therefore that these vents are near the centre of the roughly circular outcrop of the Portrush Sill (locality 1) which is known to be 45m thick and forms Ramore Head at Portrush, the islands of the Skerries several kms offshore to the north-west of the White Rocks, and also probably the rocks known as the Storks about 2 kms offshore

to the north east of the White Rocks (Fig. 30). It is known that this large body of dolerite occurs over the area at a relatively shallow depth in the crust and it would have been more than capable of supplying the explosive outgassing required to produce these explosion vents.

5.3 *Craignahulliar Quarry.*

Craignahulliar Quarry is located about 3 km to the south-east of Portrush and is just outside the area designated as the AONB. For many years it has been popular with parties of geologists who were also visiting the nearby Giant's Causeway. Since the quarry was formerly owned by a company called the Portrush Columnar

Basalt Company it should come as no surprise that the rock extracted was Causeway basalt. Quarrying has been discontinued for a number of years but there still exists some excellent examples of features of Causeway Tholeiite volcanology, some of which are unique to this locality. The site is not open to the public and anyone wishing to examine the exposures must seek authorisation from the present owners. At the time of writing it is proposed that the site will be developed as a municipal waste tip. The preservation of all the features noted below is assured and a geological trail linking them together is planned.

5.3.1 General features of the geology.

The succession exposed in Craignahulliar Quarry includes the lowest member of the Interbasaltic Formation, the laterites of the Port na Spaniagh Member (Fig. 26, section 3.8), representing the weathered and undulating surface of the Lower Basalts. This surface had been weathered and eroded forming a landscape with river valleys and a vegetation cover, before being inundated by the renewed volcanic activity of the Causeway lavas (Fig. 27, section 3.8.). Overlying the laterites, which have thin discontinuous layers of lignite in places, are several flows of the Causeway basalts, the stratigraphical equivalent of the lava

succession exposed around the Giant's Causeway. An undulating laterite surface is uncovered in places allowing visitors to walk on an ancient soil.

Locality 3a.

See map, Fig. 38. Taking the main access road into the quarry from the present office and weighbridge, a steep bank can be seen behind the old workshops and conveyor buildings. In this face can be seen the upper part of the Port na Spaniagh laterite overlain with thin pockets of blackish lignite containing abundant leaf and bark fragments and imprints (Fig. 39). This lignite has in the past yielded pollen and macro-fossils from a range of tree species including cedar, pine, spruce, hazel and alder. The first flow of the Causeway lavas erupted over this surface and a crudely developed colonnade and entablature structure can be recognised. A distinctive feature of the first flow at this locality is the presence of large gas cavities or "vesicle cylinders" which have penetrated the flow from the base upwards. These vesicles

Fig 39 The upper part of the Port na Spaniagh laterite overlain by thin layers of lignite and the first flow of the Causeway basalts.

Fig 40 Flow-foot breccia, chilled basalt in a palagonite matrix.

lent example of flow-foot breccias formed when lava flows into standing water and forms hyaloclastite deposits (see section 3.6, describing the examples exposed at Port na Spaniagh). Fig. 40 shows a 2-3m thick deposit of rounded pillow-like masses of chilled basalt in a matrix of yellow /orange coloured palagonite. The basalt flow overlying this deposit is in turn overlain by a thin bed of fine sandstone or siltstone containing plant remains and probably representing a lake-bed deposit (locality 3b(ii) and Fig. 41). These lakes may have been formed by the disruption of the river drainage pattern caused by the eruption of the lavas. This sediment buried and thus preserved the former glassy top of the underlying lava

may represent gases released by the inter-action of the hot lava with the underlying vegetation and groundwater. They are larger than vesicles normally found in the Antrim lavas, 5-10 cm in diameter and up to 20 cm in length. A further unusual fea-ture is that they are commonly filled with the mineral chalcedony, which is a form of quartz very similar to flint, rather than with zeolites which are the more common cavity in filling.

Locality 3b.

The access road south-east in-to the quarry from locality 3a climbs to a higher level with the first flow of the Causeway Tholeiite Member again exposed over-lying the Inter-basaltic laterites. At locality 3b(i) there is an excel-

Fig 41 Lava flow top under fine-grained siltstone.

Fig 42 "Chisel marks" in Causeway basalts.

known as "chisel marks". The precise origin of these markings is still under discussion but they almost certainly represent stages in the gradual growth of the colonnade columns as the interior of the lava cooled (Fig. 42).

5.4 The Giant's Causeway.

This excursion covers the geology of that part of the AONB from the Visitors' Centre at the Giant's Causeway to Port Reostan (see locality map, Fig. 43). Car parking and other facilities are available at the Causeway Visitors Centre. The main geological features in this area are the Lower Basalts, the laterites and the Causeway Tholeiites of the Interbasaltic Formation (Fig. 26 gives the summary stratigraphy, section 3.8). The lower path is now closed and inaccessible to the west of Port na Spaniagh and no attempt should be made to pass beyond the barrier. Visitors are reminded that the Causeway is a World Heritage Site of special significance to geological science and as such the area is a "no-hammer" zone and the taking

flow which is now exposed by erosion (Fig. 41). This shows a flow top marked by small domes and blisters. Visitors are asked not to disturb materials at this locality. The blisters and associated sediments are very fragile..

Locality 3c.

To the east of this section, in an older and more overgrown part of the quarry, there are very good examples of regular colonnade columns. These particular columns show the regular horizontal banding

Fig 44 Laterite of the Interbasaltic Formation underlying the first flow of the Causeway basalts.

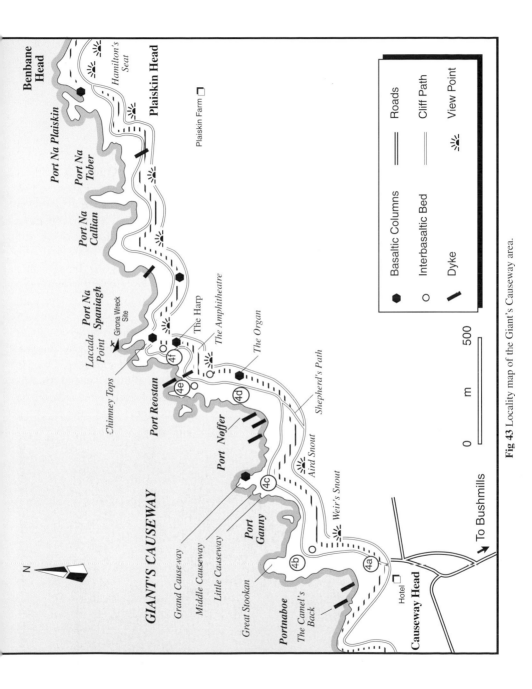

Fig 43 Locality map of the Giant's Causeway area.

Benbane Head

Plaiskin Head

Port Na Plaiskin

Hamilton's Seat

Port Na Tober

Plaiskin Farm

Port Na Calllian

Port Na Spaniagh

Lacada Point

Girona Wreck Site

The Harp

The Amphitheatre

Chimney Tops

Port Reostan

The Organ

Shepherd's Path

Port Noffer

Aird Snout

Weir's Snout

GIANT'S CAUSEWAY

Grand Causeway

Middle Causeway

Little Causeway

Great Stookan

Port Ganny

Portnaboe

The Camel's Back

Hotel

Causeway Head

To Bushmills

N

	Basaltic Columns		Roads
	Interbasaltic Bed		Cliff Path
	Dyke		View Point

0   m   500

53

The junction between the laterites and the Causeway basalts can be seen to be steeply dipping downhill towards the Causeway itself to the east and this represents one wall of the valley described earlier as having been cut into the Lower Basalt surface during the long period of dormancy that occurred at the end of the first cycle of volcanism (Fig. 20 in section 3.4).

Fig 45 The Camel's back dyke intruded into Lower Basalts.

of samples is prohibited.

Locality 4a. Descent towards Great Stookan.

The road down to the Causeway from the Visitors' Centre passes an outcrop of red lateritic material some 200m east of the building (Fig. 44). This marks the top of the Lower Basalt surface and above this is the first flow of the Causeway basalts. Seawards the Lower Basalts are exposed in the cliffs below the Causeway Hotel and on the foreshore is the sea stack known as the Camel's Back which is formed by a dyke cutting through the Lower Basalt lavas (Fig. 45).

Locality 4b. Windy Gap and Great Stookan.

As the path turns the corner eastwards (and often with less shelter from the wind,

Fig 46 Spheroidal weathering in lava flows of Lower Basalt age.

54

Fig 47 Flow junctions in Lower Basalts.

valley exposed on the cliffs on the east side of Port Noffer (Fig. 48). From this point return to the path and proceed to the Causeway.

Locality 4c. Giant's Causeway.

The Causeway (Fig. 49) is formed by the ponding of the first flow of the Causeway lavas in the bottom of the river valley carved out of the Lower Basalt surface. The Interbasaltic laterite can be seen at low tide and passes underneath the Causeway itself.

hence the name) there are exposures of Lower Basalts with a characteristic SPHEROIDAL WEATHERING. Here large blocks of basalt, produced by the jointing pattern of the lavas, have been rotted by physical and chemical weathering to form rounded boulders where the weathering has formed concentric layers of material. It is sometimes referred to as "onion-weathering" as the layers have the appearance of the internal structure of onions (Fig. 46 and 53).

From this point leave the path and walk towards the sea and the cliffs at Great Stookan.

The cliffs at the Great Stookan show a succession of thin Lower Basalt flows with characteristic red amygdaloidal tops to each flow. The junction between the bottom two flows is clearly seen near the high water mark and shows amygdales or gas bubbles filled with white zeolite minerals (Fig. 47).

This locality gives clear views of the Giant's Causeway to the east, with the Interbasaltic laterites of the other side of the former Lower Basalt river

Fig 48 The Giant's Causeway and headlands from the west.

55

Fig 49 The Giant's Causeway.

Fig 50 The colonnade forming the columns of the Grand Causeway, passing up into the entablature in the cliffs of Aird Snout.

The ponded lava cooled to form a colonnade as seen in the Little, Middle and Grand Causeways passing up into the entablature seen in the cliffs of Aird Snout on the landward side of the Causeway (Fig. 50). The near horizontal columns seen on the landward side of the Causeway near the path are probably the result of an earlier landslip, pulling the originally vertical columns over. This feature can be seen in a number of localities around the Causeway coast where there have been more recent landslips.

Looking inland from the Causeway it is possible to appreciate the width and slopes of the sides of the former river valley. The Interbasaltic can be clearly seen at a higher level both to the east and west and is well exposed to the east at Port Noffer, with several flows of the Lower Basalts below the laterites and two flows of the

Fig 51 The basalt sequence at Port Noffer.

Flow 2

Grassy ledge between
flow 1 & flow 2

Entablature

Flow 1

Colonnade

Giant's Organ

Fig 52 The flow structure at the Giant's Organ.

Causeway basalts above (Fig. 51) and the cross section in Fig. 20, section 3.5). From the Grand Causeway follow the path around Port Noffer to the Giant's Organ and stop at the junction with the Shepherd's Path.

Locality 4d. Giant's Organ, Port Noffer.

The Giant's Organ (Fig. 52) is the colonnade of the first flow and as such is the same part of the flow as the Giant's Causeway. Here however, only the sides of the columns can be seen, and not the sides and tops as at the Causeway. The change from colonnade jointing to entablature jointing is very sharply defined and the differences in width and style of columns are clearly evident. Above this first flow the second flow can be seen, separated by a grassy ledge which has formed along the softer and therefore more easily weathered top of flow 1.

Locality 4e. Port Noffer to Port Reostan.

The path is now at the level of the Port na Spaniagh laterite of the Interbasaltic Formation (Fig. 26) and the reddened altered Lower Basalt surface is clearly visible, up to 5m thick at this point. Beside the path are good examples of the "Giant's Eyes" (Fig. 53). These are residual masses of more resistant basalt which are less chemically altered than the lateritized

Fig 53 The "Giant's Eyes" residual basalt within the weathered laterite.

material around them. The spheroidal weathering noted previously at Windy Gap (locality 4b) represents a much earlier stage in the weathering process but the similarities in form are obvious. On the headland between Port Noffer and Port

Reostan a dyke can be seen at sea-level cutting through the Lower Basalts (Fig. 54). This can be traced upwards to where it passes through the laterites and splits into several smaller sheets in the softer weathered material.

Fig 54 Dyke cutting through lava flows at Port Reostan.

Locality 4f. Port Reostan.

The feature known as the "Harp" (Fig. 55) is formed of the slightly curved colonnade columns of the first flow. This curvature may be due to distortion of the cooling surfaces by slow movement in the partly solidified flow, leading to a change in the angle of growth of the columns. The landmark known as the Chimney Tops are the isolated columns of the colonnade of the second flow. The "Amphitheatre" in Port Reostan shows flow 2 without an entablature, with flow 3 showing a thin development of both colonnade and entablature at the top of the cliff (Fig. 56).

5.5 White Park Bay to Ballintoy.

The section of the north Antrim coast from White Park Bay eastwards to Ballintoy Harbour, a distance of about 2kms, contains almost the whole range of rock types present in the AONB (see geological map, Fig. 4 and stratigraphical column Fig. 5, section 1.3). Parallel to the coast in this area is a major break or fault in the earth's crust, the Portbraddan Fault, and it is estimated that north of this fault line the crust has moved downwards by about 100m. This has had the effect of bringing the Upper Basalts down to the same level as the Cretaceous Chalk at sea level on the west side of Whitepark Bay at Portbraddan. Elsewhere both the Lower and Upper Basalts are above the Chalk.

5.5.1 Ballintoy Harbour.

Turn off the main Portrush to Ballycastle road on to the side road sign-posted to Ballintoy Harbour and follow this road to the harbour and car park where toilets are available. Figure 57 shows the geology of Ballintoy harbour. Localities 5a, 5b, 5c and 5d are accessed from the car park walking west along the coastal path.

To examine locality 5e, retrace your steps to the harbour and continue east along the beach.

From the eastern end of White Park Bay towards Ballintoy Harbour there is a contrast in the surface geology across the line of the Ballintoy Fault which is an eastward extension of the Portbraddan Fault (see Fig 57). To the south of the Ballintoy Fault line the Cretaceous Chalk is found at the surface, forming the raised beach cliffs behind the present shore line, while to the north and on the small off-shore islands are various members of the Antrim Lava Group.

A cross-section through the fault from north to south shows that the northern block has moved down relative to the southern side, so that north of the fault the Chalk is at some depth below the surface. This is in contrast to its occurrence at sea-level on the south side (see cross-section Fig. 57). The vertical movement along this fault is probably at least 100m and rather than being a sin-

Fig 55 The Giant's Harp.

Fig 56 The lava sequence in the "Amphitheatre" in Port Reostan.

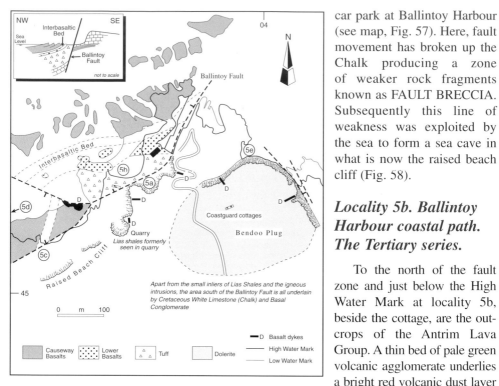

Fig 57 Geological map of Ballintoy Harbour (after Wilson and Manning 1978).

Map labels:
NW — SE
Sea Level
Interbasaltic Bed
Ballintoy Fault
not to scale

04
N
Ballintoy Fault
Interbasaltic Bed
(5e)
(5h)
(5a)
D
D
D
D
D
(5d)
D
Coastguard cottages
Bendoo Plug
Quarry
Lias shales formerly seen in quarry
(5c)
Raised Beach Cliff
45

Apart from the small inliers of Lias Shales and the igneous intrusions, the area south of the Ballintoy Fault is all underlain by Cretaceous White Limestone (Chalk) and Basal Conglomerate

0   m   100

Causeway Basalts | Lower Basalts | Tuff | Dolerite
D Basalt dykes
—— High Water Mark
—— Low Water Mark

car park at Ballintoy Harbour (see map, Fig. 57). Here, fault movement has broken up the Chalk producing a zone of weaker rock fragments known as FAULT BRECCIA. Subsequently this line of weakness was exploited by the sea to form a sea cave in what is now the raised beach cliff (Fig. 58).

Locality 5b. Ballintoy Harbour coastal path. The Tertiary series.

To the north of the fault zone and just below the High Water Mark at locality 5b, beside the cottage, are the outcrops of the Antrim Lava Group. A thin bed of pale green volcanic agglomerate underlies a bright red volcanic dust layer under a few thin flows of Lower Basalts. These are in

gle line of weakness it is probably several faults which are roughly parallel and forming a fault zone a few tens of metres wide.

Locality 5a. Ballintoy Harbour car park.

The effect of one of these faults on the Chalk can be examined at locality 5a near

Fault

Fig 58 Former sea cave caused by erosion along fault breccia formed along the Ballintoy fault.

Causeway basalt

Lower basalt

Red volcanic dust

Volcanic agglomerate

Fig 59 Lower basalts underlain by red volcanic dust with Causeway basalts to seaward.

towards White Park Bay. This particular part of the north Antrim coastline between Ballintoy Harbour and Whitepark Bay shows very clearly the effects of the sea-level changes that occurred after the last retreat of the great ice caps some 15,000 years ago. The enormous thickness of ice cover during the Ice Age had caused the crust to sink and recovery from this sinking took longer than the time required for the ice caps to melt. The rapid rise in sea level caused by the melting ice meant therefore that much of this depressed landsurface was temporarily covered by the sea. As the land slowly

turn overlain by lavas of the Causeway basalts which show their characteristic jointing pattern and form the off-shore islands at this point (see map, Fig 57. and Fig. 59). Between these two sets of lavas are the Interbasaltic laterites, but these are not exposed above sea level and their position is inferred on the geological map. The present harbour at Ballintoy is approximately along the line of the fault zone, with basaltic rocks to the north-western side and Chalk to the south-east (Fig. 60).

Locality 5c. Ballintoy Harbour coastal path. The raised beach features.

The track continues westward

Basalt

Rathlin Island

Chalk

Ballintoy fault

Downthrow side

Upthrow side

Fig 60 The Ballintoy Fault along the line of Ballintoy Harbour.

61

The lowest beds of the Chalk exposed just above the High Water Mark contain numerous pebbles and rock fragments and because of their large grainsize they are referred to as CON-GLOMERATES. These coarse grained beds are the result of a gradual flooding of the Jurassic land surface by the sea dur-

Fig 61 Sea stacks and arches on the raised beach west of Ballintoy Harbour.

emerged again this former coastline can now be seen as a raised beach some 5-10m above the present High Water Mark. It is marked by a line of former sea cliffs behind the present coast (see map, Fig. 57), often with caves (for example locality 5a), and more obviously as sea stacks and sea arches which are now left well above the present water level (Fig. 61), seen approximately 400m west of the carpark at Ballintoy Harbour.

5.5.2. White Park Bay.

Locality 5d. White Park Bay eastern end.

On the east side of White Park Bay (see map, Fig. 62) the base of the Cretaceous Chalk is seen sitting above the Lias Clay, which is exposed just below the High Water Mark, sand levels permitting. As the Lias is a sub-division of the Jurassic Period (see stratigraphic column, Fig. 5), this junction therefore represents a major break in the time sequence, or UNCONFORMITY, and also indicates a marked change in sedimentary conditions.

ing Cretaceous times. The conglomerates also have a distinctive speckled appearance, occasionally greenish, due to the presence of the mineral glauconite (see section 5.2, locality 2).

The Lias clay formed as an offshore marine deposit and is a sticky grey clay with occasional harder dark limestone

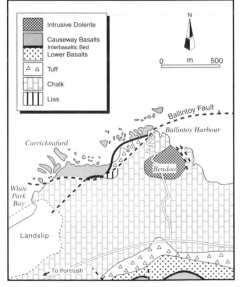

Fig 62 Geological map of Whitepark Bay-Ballintoy Harbour (after Wilson and Manning, 1978).

62

possible to recognise radial contraction joints within the plug that formed at right angles to the main cooling surface. The vertical contact with the Chalk is clearly visible in the cliffs (Fig. 64). When hot magma comes into contact with the colder rock it is intruding, there is a baking effect similar in many

Fig 63 Line of Portbraddan-Ballintoy Fault across White Park Bay.

bands and abundant marine fossils such as ammonites, LAMELLIBRANCHS, CRINOIDS and sea-urchins. The hummocky ground behind the beach at White Park Bay is the result of slipping of the overlying Chalk cliffs on this weak Lias clay, similar to the process that produces the instability of the Coast Road in east Antrim.

The Portbraddan – Ballintoy Fault crosses the coastline here on the east side of White Park Bay (see map, Fig. 62) with brecciated Chalk forming seastacks on the shore side and basaltic lavas forming the off-shore islands (Fig. 63). The fault crosses the coastline again on the west side of White Park Bay and in the middle distance in Fig. 63 the contact between Chalk and down-faulted basalt at Portbraddan can be seen.

ways to the effects of firing clay pots in a furnace. The heating by the magma often hardens and discolours the contact rock and these changes are known as CONTACT or THERMAL METAMORPHISM, where

Fig 64 Contact zone between the dolerite of the Bendoo Plug and the Chalk.

Locality 5e. Bendoo Plug.

Return eastwards along the track from White Park Bay to the car park at Ballintoy. Just east of Ballintoy Harbour is the Bendoo PLUG (see map, Fig. 57). This is formed of a medium grained basaltic rock known as dolerite and the plug is roughly cylindrical in form with a diameter of about 350m. It is

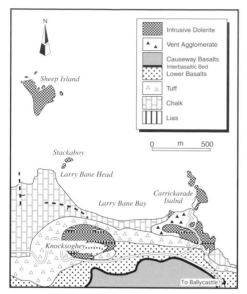

Fig 65 Geological map of Larrybane Bay
and Carrickarade Island
(after Wilson and Manning, 1978).

Fig 66 The Knocksoghy Sill at Larrybane Bay.

5.6 Carrickarade.

Locality 6. Carrickarade.

East of Ballintoy Village, on the Ballycastle road, a signposted road to Larrybane leads to a car park with additional facilities available. Walk east along the cliff top path for Carrickarade. The main features of the volcano can be seen from Carrickarade Island, which can be reached by crossing the rope bridge.

The spelling of Carrickarade adopted throughout this book has been used consistently in the geological literature over many years. Carrickarede is the more usual spelling.

metamorphism is the term for changes brought about by increases in temperature or pressure. The observed changes at the contact between the plug and the Chalk at Ballintoy are relatively slight, the zone of metamorphism is only a few centimetres wide. This suggests that the plug was a relatively short-lived intrusion and was not a major feeder for the Antrim lavas, as for example Slemish was. Similar cylindrical features, termed "pit-craters", are known from active volcanic areas such as Hawaii. They are caused by the crust subsiding during lava eruption and they form temporary lava lakes.

Fig 67 Sheep Island and Larrybane Head.

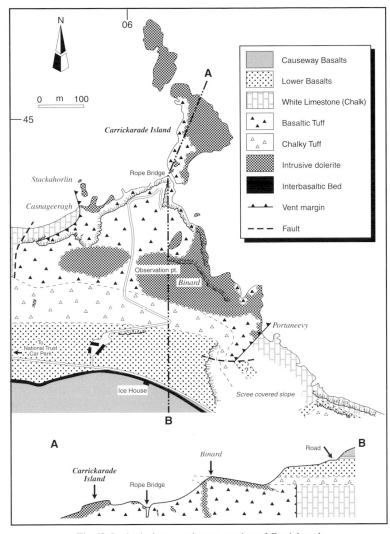

Fig 68 Geological map and cross-section of Carrickarade
(after Wilson and Manning, 1978)

dolerite of the Knocksoghy Sill, one of the few sills in Antrim intruded into the lavas, part of which probably forms Sheep Island offshore. (See geological map, Fig. 65 and Figs. 66 and 67). Fig. 65 shows the extent of the Carrickarade volcano, almost 600m from east to west and approximately 600m from north to south around Carrickarade Island. The volcanic remains consist of exposures of agglomerate made up of fragments of basalt, Chalk and Lias clay, blown out from a volcano which existed during the initial explosive phase of the igneous activity in Antrim. The details of the geol-

While the island of Carrickarade and its rope bridge are well known features to many visitors to the north coast, what is perhaps not so well known is that the island and the area around it on the mainland are the remains of an explosive volcano which is unique in its extent in the whole of the Causeway Coast AONB. The disused quarry in the hillside behind the National Trust Visitor Centre shows the

ogy and a cross-section are shown in Fig. 68. The cliffs on the east side of the island are part of a later dolerite intrusion but the rock below the bridge and on the mainland is basaltic agglomerate. This is the volcanic opening or vent and is filled with an agglomerate consisting of blocks of basalt up to 3m in diameter embedded in a matrix of volcanic ash, with occasional fragments of the older rocks in the area such as the

Fig 69 Solution hollows on Carrickarade Island formed by weathering of the limestone component of the agglomerate.

the main part of the island to which the rope bridge is attached, and the smaller island beyond it, Carrickarade Island (Fig. 68). The principal features of the volcanic complex can be seen from the summit of the island with reference to this geological map, and as the cliffs here are extremely dangerous no attempt should be made to approach the edge. At Binard, along the coast to the south-east of the rope bridge, most of the cliff consists of a dolerite intrusion, with smaller intrusions around Stackahorlin, south-west of the bridge. At Portaneevy, further to the east, the contact is near vertical with both agglomerate and dolerite against the Chalk, although the

Lias clay and the Chalk, through which the volcano was erupting. Much of the ash shows hollows and cavities on the surface which suggests there may be a finely ground limestone component which is being dissolved out more easily by weathering processes. Fig. 69 shows Carrickarade from the south with the characteristic solution hollows.

The agglomerate in the vent is intruded by a number of veins and irregular masses of dolerite which were part of a later phase of the history of the volcano when explosive activity had waned and magma, rather than agglomerate, was the main product. One of these intrusions forms

Fig 70 The western vent margin of the Carrickarade vent against Chalk and Lower Basalts.

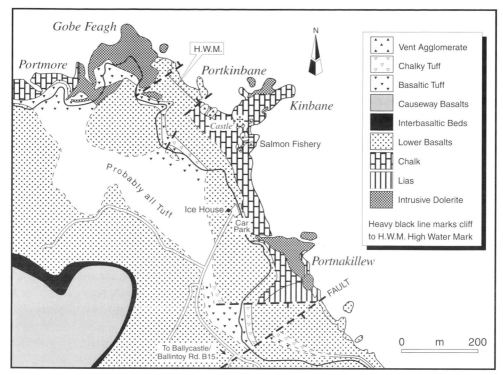

Fig 71 Geological map of the Kinbane Head area (after Wilson and Manning, 1978).

margin itself is obscured by scree deposits. The western margin (Fig. 70) can be seen to the west of the small island of Stackahorlin where it rises steeply in the lower part of the cliff and then flattens out to the west overlying the Chalk at the cliff top. Above the Chalk there is a layer of chalky tuff 25m thick overlain by lavas of the Lower Basalts and Causeway basalts (see the cross-section with geological map, Fig. 68). It is thought likely that this tuff layer was erupted from vents in the Larry Bane Head area about 1 km to the west.

Walking west from the car park down a stone track leads to the main quarry of Larrybane. There is excellent exposure through the Chalk with a number of hard-grounds evident. Continue down the track to the beach then walk eastwards towards Carrickarade. A small cave can be seen in

the cliff with some SPELEOTHEM development comprising a pillar and STALAC-TITES. Such features are very unusual in the Cretaceous Chalk and visitors are asked not to damage them.

5.7 Kinbane Head.

Locality 7. Kinbane Head.

Kinbane Head lies some 3 kms to the east of Carrickarade. Access is by the turn-off from the Ballycastle to Ballintoy road (B15) at the signpost to the Kinbane Castle and then proceed to the car park on the cliff top.

This site appears to be a further example of a partially preserved volcanic vent associated with the early explosive phase of the volcanic activity in Antrim. Along the top of the basalt cliffs, viewed from the car park behind Kinbane Head, can be

Basaltic tuff

Chalky tuff

Lower Basalts

Fig 72 View of the succession at Kinbane showing Lower basalts overlain by chalky and basaltic tuffs.

seen thick tuff and agglomerate beds which get thinner in the direction of Ballycastle. Take the path down to Kinbane Castle in Portkinbane. In the cliffs here thick deposits of chalky tuff overly the Lower Basalts and are in turn overlain by a layer of basaltic tuff. The general distribution of the rock types is shown on the geological map, Fig. 71 and also on Fig. 72.

The Chalk at the base of the succession can be seen below the ruins of Kinbane Castle and has been broken and shattered by intrusions of dolerite and basaltic agglomerate (Fig. 73). To the north of Kinbane Castle sea erosion has uncovered the contact zone of a dolerite intrusion into Chalk. The forceful nature of the intrusion can be shown by the shattered state of the Chalk overlying the dolerite.

Fig 73 Chalk beneath Kinbane Castle shattered by the effects of igneous intrusion.

6

Geology and Scenery on the Causeway Coast

6. Geology and Scenery on the Causeway Coast

The overall appearance of any landscape is the result of the combination of a number of factors of which bedrock geology, structures such as faults, weathering and erosion are probably the most important. Scenery is defined as views or prospects which are considered to be beautiful, picturesque or impressive. In the case of the Causeway coast, which is generally regarded as falling into this category of a scenic landscape, the bedrock geology consists of a relatively small number of rock types. These, as we have seen, are the basalt lavas and associated intrusions, the Chalk, and the Lias clay. It is the disposition of these rocks and the effect of weathering and erosion over millions of years, together with vegetation and the effects of human activity, that have combined to produce the present day landscape.

In north Antrim, the most dramatic views are those along the coast resulting from active erosion by the sea. This has been the most important erosion agent for millions of years but particularly so since the end of the last glacial period. The mainly east - west line of the north Antrim coast is defined largely by the major faults trending in this direction such as the Portbraddan Fault. Cutting these faults at almost right angles are additional faults such as the Portballintrae Fault (see geological map, Fig. 4, Section 1.3). These produce zones of weakness along the coast which are exploited by marine erosion and represent an important factor influencing the positions of the main bays. (Fig.74). The present coastline with its prominent basaltic and Chalk cliffs, is the result of an intense period of erosion following ice withdrawal. Large blocks of rock would have been prised from the cliff faces by severe frost action as Northern Europe would still have been experiencing a bitterly cold climate. This process would have steepened the cliffs and formed extensive screes at the base. As the vast ice sheets in the northern hemisphere melted the level of the sea rose more rapidly than the land, which was also rising, having been relieved of its load of ice. This time lag in the rise of the land meant that the sea had time to cut a shoreline above its present level, and this raised beach or shoreline is preserved as sea cliffs, stacks and arches preserved above the current High Water Mark (Fig. 61), as described in section 4.2.

The surface geology of the Causeway coast is dominated by the Antrim lavas

Fig 74 Portballintrae Bay developed on the Portballintrae Fault.

which, in their final stages of eruption, must have covered a substantially larger area to the north and west than present outcrop. The plateau must certainly have been considerably thicker as it is known from younger lava successions, such as in Iceland, that there is a zeolite free zone in the top 700-1000m of basalt successions. This zone is missing in Antrim and lavas here contain zeolite minerals in the amygdales throughout the succession, so that some 700m of basalt has been removed by erosion during the last 55 million years or so. In the period immediately following a major lava plateau building episode there is always a good deal of crustal readjustment due to settlement after the eruption of such vast quantities of magma from depth. Such a readjustment took place in Antrim in the form of gentle flexing or bending to give a general synclinal form to the plateau and a considerable number of faults. Some of these faults had previously existed and were simply re-activated while others represented new structural features. This faulting has resulted in large blocks of the countryside moving up, down or laterally relative to each other along the fault planes. A section or slice through the north coast (Fig. 4) shows that the various layers represented by the Antrim basalts, the Chalk and the Lias clay appear at different levels relative to each other along the coast. This effect is most clearly seen at White Park Bay and Ballintoy Harbour where basalts have been brought down to

sea level by movement along the Port Braddan Fault and are seen side by side with the Chalk. (Fig. 75). Faulting does not change the order or succession of these rock types, the basalts are always younger and therefore overlying the Chalk, but they do appear at different levels along the coastline.

Fig 75 Port Braddan Fault brings basalt to the same level as the chalk.

Where the coast is formed by basalt flows the shoreline tends to consist of steep cliffs, as exemplified the cliff path to the east of the Giant's Causeway itself. The softer lateritized flow tops can be more easily eroded by the sea and this undermines the cliff which falls away to produce a near-vertical cliff face. The relative softness of the Interbasaltic laterite bed is one of the main difficulties in keeping the coastal cliff path around the Causeway Coast open, as it is frequently worn away and causes rock falls which destroy the path or block it. This produces a two tiered effect on the cliffs eastward of the Causeway, with steeper slopes below and above the laterite (Fig. 51, in Section 5.4. The occurrence of the Interbasaltic horizon at sea level on either side of the Giant's Causeway has enabled the sea to erode the softer laterite and thus form the promontory which is the Giant's Causeway. Dykes are often more resistant than the basalt lavas they intrude and the dyke on the headland between Port Noffer and Port Reostan stands out as a hard resistant wall right down to sea level (Fig. 54).

Other intrusive rocks which are hard and resist erosion can be seen at Ramore Head where the Portrush Sill forms a prominent headland (Fig. 39) and the steep cliff formed by the Knocksoghy sill near Carrickarade (Fig. 66). At Carrickarade the ropebridge crosses the narrow chasm cut in the soft agglomerate material of the vent, while to the north and south of this there are hard dolerite intrusions forming the offshore island and the mainland to the south. At sea level, the Chalk headland of Kinbane Head consists of dolerite intruded into the Chalk and is a further example of resistant intrusive rocks forming promi nent features.

Where Chalk occurs at the coast it also frequently forms steep or vertical cliffs as at the White Rocks or Larrybane Head, but where the underlying soft Lias clay is at or near the surface, the limestone can be undercut and the coastline eroded back much more easily. This is well demonstrated at White Park Bay where the pres- ence of Lias clay at sea level has under-mined the Chalk cliff which has slumped to produce a spectacular bay backed by hummocky ground in front of a low Chalk cliff 400m to the south. Similarly at Bushfoot Strand, just west of the Giant's Causeway, a faulted block of Chalk between headlands of basalt has enabled the sea to erode a bay into which has drift-ed sands carried by the prevailing coastal or longshore drift from the west. This has formed the sandy beach and the extensive area of windblown dune sand behind it.

Portballintrae is a good example of a line of crustal weakness along a fault, in this case the Portballintrae Fault, being exploited by marine erosion to produce a small inlet forming a natural harbour.

South from the coastal area the land-scape consists of low and irregular hills, sometimes showing signs of the layering associated with lava flow sequences, and generally dipping southwards. The surface geology away from the coast is almost entirely basaltic and is mostly covered by a layer of boulder clay deposited by the ice sheets during the Ice Age.

7

Human Activity on the Causeway Coast

7. Human Activity on the Causeway Coast

7.1 Early Settlement.

One of the influences on the form of the landscape is the settlement pattern of those who live on the land or who use it. The earliest human inhabitants of the Causeway Coast area were the MESOLITHIC hunters and fishermen who lived mainly in coastal localities, but also followed the Bann Valley southward to the Lough Neagh lowlands. They arrived in Ireland some 9,000 years ago.

The presence of large amounts of flint in the Chalk cliffs around north Antrim was an important attraction for the migration of people into the area and evidence of flint-mining and flint tool manufacture represents one of the earliest known examples of natural resource exploitation in Ireland. It is interesting in this context to note that the Antrim Chalk outcrops are the only extensive outcrops of flint-bearing rocks for many hundreds of kilometres in any direction and so provided an important raw material for exploitation locally. The physical nature of the coastline had a major influence on the settlement pattern and the high cliffs and rugged promontories associated with the Chalk and basalt outcrops would have been avoided in preference for the sandy bays backed by sand

dunes such as at the mouth of the Bann or Bush Rivers, or at White Park Bay.

The best known Mesolithic site in this area is at Mount Sandel near Coleraine on the River Bann. This was first occupied about 9,000 years ago and the large quantities of flint implements and other materials found at the site provides a picture of hunter-gatherers who lived on fish such as salmon, eels and trout, who hunted animals such as wild pig and who collected and stored hazel nuts and other plant foods.

An improvement in climate after the Ice Age ultimately allowed trees such as oak, elm, hazel and pine to flourish and by 6,000 years ago much of Ireland, including the upland areas, was covered in thick forest. The NEOLITHIC farmers who succeeded the Mesolithic hunter-gatherers around 6,000 years ago were the first inhabitants to make a substantial impact on the landscape of north-east Ireland. To create space for their crops and livestock they began to selectively clear the forests, and as metal technology developed by the BRONZE AGE, this clearing continued with increasing efficiency. Deforestation,

Fig 76 Neolithic monument at Clegnagh near Portrush.

coupled with a cooling climate in the last few thousand years, has resulted in a great reduction in the area of the island covered by trees. Much of this previously forested area is now covered in peat bog. The roots and stumps buried as the bog grew are frequently re-exposed as the peat is removed by digging, a further example of modification of the landscape by human activities.

Settlement sites of Neolithic farmers within the AONB are known from the coastal sand dunes of the White Rocks at Portrush and at White Park Bay. A feature of Neolithic settlement is the occurrence of tombs of various types, usually constructed from large stones and referred to as MEGALITHS. (Fig.76). Evidence for the use of metal for tools and weapons in Ireland goes back to about 4,500 years ago and County Antrim is especially rich in such remains with, for example, a Bronze Age casting workshop at White Park Bay.

It is generally considered by archaeologists that the third stage of technological development, following the use of stone and bronze implements, was the use of iron. Earthen-banked and ditched raths, stone-walled cashels or the timber-built crannogs found on islands in lakes are characteristic of this period in Ireland. Within the AONB there are about twenty such ring-works including five cashels extending in an east-west line from

Ballintoy Bushmills. There is evidence of Iron Age settlement at Mount Sandel to the west of the area and an Iron Age hearth has been found at the White Rocks.

The stone castles of a later age at Dunluce (Fig. 77) and Dunseverick are both likely to have replaced earlier earthenbanked promontory forts which utilised sea cliffs for defensive purposes. There is evidence that forts on sites such as Dunseverick were associated with minor kingdoms which existed in the early Christian era, post 300 AD.

Fig 77 Aerial view of Dunluce Castle.

Following the Norman conquest of England and parts of Ireland a number of fortifications were erected in Ulster, mostly in coastal areas, including Dunluce Castle, built probably some time in the 14th century.

The remains of fortifications presently seen at Kinbane and Dunseverick date from the 16th century and reflect the instability associated with the struggle for power between the various ruling families of the area such as the O'Neills, the O'Donnells and the McQuillans. The Plantation of Ulster at the beginning

Fig 78 Old adit mine, Port Moon near Dunseverick.

no further details known of its occurrence or extent. A thin bed of lignite occurs between flows 1 and 2 at the Giants' Causeway beside the steps of the Shepherd's Path which leads from the Giant's Organ to the cliff top behind Port Noffer. A thick bed of lignite was discovered in Craignahulliar Quarry just south of Portrush, which was mined and sold for fuel in the area during World War 2.

Mining of the Interbasaltic laterites for iron ore and bauxite (aluminium ore) was more extensive and began about 1860 in mid-Antrim with the first output from north Antrim mines recorded in the mid-1880's. The chemical processes involved in the formation of laterite from basalt tend to reduce the silica content of the basalt along with other elements such as calcium and sodium, leaving a greater concentration of iron and aluminium, generally in the form of oxides and hydroxides. It is this greater concentration of aluminium and iron in the residual laterite that meant it could be mined as a raw material for use as a flux in the iron furnaces of Cumberland when it was iron-rich or, later on, in the manufacture of aluminium sulphate (alum) when aluminium-rich. Most of the mines in north Antrim were concerned with the production of bauxite; for example mines near White Park Bay which closed around the period of the First World War. Mines in the west of the area, such as those at

of the 17th century introduced the next phase of settlement to north Antrim, this time mainly from Scotland. This influx allowed the development of new businesses and industries, including the distillery at Bushmills in 1609, still active on the same site today. The growth of the tourist industry from the late 19th century onwards meant that small villages such as Portrush, Portballintrae and Ballintoy expanded as centres for visitor accommodation, and more recently as dormitory areas for the larger industrial, administrative and educational centre that Coleraine has grown into.

7.2 Mining and Quarrying.

Mining in this area has been recorded from the mid-eighteenth century when a bed of lignite, within the Interbasaltic laterites at Ballintoy, was worked. Traces of several old ADITS or horizontal shafts, can be seen on the hillside above the village, together with a series of spoil heaps. Lignite was also worked around this time at the Giant's Causeway where the bed was said to be up to 2m thick, but there are

Fig 79 Aerial view of the former limestone quarry at Larrybane Head.

Craignahulliar, south of Portrush, probably finished producing bauxite ore towards the end of the last century. Some of these old mines are still open, they were all adits driven horizontally into the hillside but they should not be entered due to the danger of collapse. (Fig. 78).

The main products of the extractive industries in north Antrim were from the many quarries producing basalt and limestone, but at the time of writing none of these are in operation. The Chalk was quarried on a large scale at localities such as Ballintoy and Larrybane Head (Fig. 79), much of it being burnt for agricultural lime production. The lime-stone was burnt in lime kilns which are quite commonly found throughout the area, for example, the harbour at Ballintoy was built to transport the lime burnt in the kilns near the quay (Fig. 80). The output of the numerous basalt quarries was principally for road metal and concrete making.

Fig 80 Disused limekiln at Ballintoy Harbour.

Appendices

APPENDIX 1. GLOSSARY OF GEOLOGICAL TERMS

ADIT - A horizontal or near-horizontal passage by which a mine is entered.

AGGLOMERATE - A rock composed of coarse fragmental debris produced by explosive volcanic activity.

AMMONITE - One of a large group of extinct sea animals related to the modern nautilus.

AMYGDALE - A gas cavity or vesicle in volcanic rocks which has become filled with secondary minerals for example zeolites.

AMYGDALOIDAL LAVA - Lava with numerous small gas-cavities, usually filled with minerals such as zeolites.

ASH - Material blown out of an explosive volcano. When consolidated into rock it is called tuff.

ASTHENOSPHERE - The layer within the Earth below the lithosphere, a relatively weak zone characterised by low seismic velocities, and probably partially molten.

BASALT - Dark-coloured fine-grained basis igneous rock.

BAUXITE - A lateritic rock rich in hydrated aluminium oxides. Used as an ore of aluminium and a source of aluminium compounds, particularly aluminium sulphate.

BELEMNITE - An extinct type of cephalopod or squid-like creature known from cigar-shaped fossils found in the Cretaceous Chalk.

BOREAL - Northern, a biogeographical region consisting of the northern and mountainous regions of the northern hemisphere.

BOULDER CLAY - Ground-moraine produced by the grinding action at the base of a moving ice-sheet. Consists of a stiff clay containing stones and boulders of all sizes.

BRECCIA - A rock composed of angular fragments embedded in a matrix, produced by the breaking up of a consolidated rock by some natural agency such as faulting, collapse, volcanic activity.

BRONZE AGE - The period in archaeology when bronze tools weapons and ornaments were produced, from 2500-3000 BC in Ulster.

CARBONIFEROUS - Period of the stratigraphic column after the Devonian and before the Permian, from approximately 345 million years to 280 million years ago.

CENTRAL VOLCANO - Volcanic eruption where the erupted materials are emitted from a central pipe or vent resulting usually in the formation of a volcanic cone

COCCOLITH - A type of marine algae with a calcareous shell.

COLONNADE - The zone of regular columns at the base of a multi-tiered columnar lava of Causeway Tholeiite type. From the architectural term referring to a range of columns placed at regular intervals and supporting the entablature.

COLUMNAR JOINTING - The division of an igneous rock into columns by cracks produced as a result of thermal contraction or shrinkage on cooling.

CONCHOIDAL FRACTURE - A type of rock or mineral fracture which produces a smoothly curved, shell-like surface. A characteristic of flint for example.

CONGLOMERATE - A sedimentary rock composed of rounded fragments which are of gravel or pebble size.

CONSTRUCTIVE PLATE MARGIN - In plate tectonics a plate margin where plates are diverging or moving apart and the crust is under tension with new crust being created.

CONTACT METAMORPHISM - The changes in the mineralogy and texture of a rock produced by contact with intrusive or extrusive igneous rocks.

CONTINENTAL DRIFT - The horizontal displacement or rotation of continents relative to one another, an essential process in plate tectonics.

CONVECTION - A mechanism of heat transfer in a flowing material in which hot material from the bottom rises because of its lower density while cool surface material sinks.

CONVECTION CELL - A single closed flow circuit of rising warm material and sinking cold material.

CORE - The central part of the Earth below a depth of 2900 km. Thought to consist of iron and nickel and to be molten on the outside with a solid inner section.

CRETACEOUS - Period of the stratigraphic column after the Jurassic and before the Tertiary, from approximately 145 million years to 65 million years ago.

CRINOIDS - A group of marine fossils related to the echinoderms consisting of an upper part with radiating arms raised above the sea floor by an elongated jointed stem.

CRUST - The outermost layer of the lithosphere. The continental crust is mostly granite in composition and the oceanic crust is composed mainly of basaltic rocks.

DESTRUCTIVE PLATE MARGIN - In plate tectonics a plate margin where plates are converging or moving together and the crust is under compression and therefore being shortened.

DEVENSIAN - The latest cold stage in the upper Pleistocene in Britain, from 50 000 to 10 000 years ago. Corresponds to the Midlandian Cold Stage in Ireland.

DEVONIAN - Period of the Stratigraphic Column after the Silurian and before the Carboniferous, from approximately 395 million years to 345 million years ago.

DOLERITE - Medium-grained dark basic rock, of the same composition as basalt but with coarser crystals.

DRIFT - A collective term for all the rock, sand and clay that has been transported and deposited by a glacier, or by water derived from ice melting.

DRIFT MAP - A geological map showing the distribution of the drift or superficial deposits.

DRUMLIN - A streamlined hill composed of glacial till with its long axis parallel to the direction of ice flow.

DYKE - Vertical or near-vertical wall-like sheet of intrusive igneous rock.

ENTABLATURE - The zone of narrow often curved columns above the colonnade columns in a multi-tiered columnar lava of Causeway Tholeiite type. From the architectural term referring to that part of the building which rests on the colonnade and includes the frieze.

ERRATIC - Rock fragment which has been transported by ice from its source and usually differs from the bedrock in the area where it is found.

ESKER - Sand and gravel in the form of a narrow ridge deposited by meltwater channels from a glacier or ice cap.

EXTRUSIVE IGNEOUS ROCK - Material resulting from volcanic activity at the earth's surface, generally solidified lava.

FAULT - A fracture or zone of fractures along which differential movement of the wall-rocks has taken place.

FAULT BRECCIA - A rock composed of coarse angular fragments resulting from crushing and grinding along faults.

FISSURE ERUPTION - Volcanic eruption from a linear fissure or crack in the earth's surface, usually with the emission of basic lava.

FLINT - A form of microscopically crystalline silica occurring in nodules in the Cretaceous Chalk (White Limestone).

FLOW-FOOT BRECCIA - Deposits consisting of hyaloclastite fragments and pillow lavas, formed when basaltic lavas flow from land into water.

FORMATION - In stratigraphy the primary unit consisting of a succession of rocks useful for mapping or description e.g. the Lower Basalt Formation.

FRACTURE - (1) the manner of breaking and the appearance of a mineral when broken, e.g. conchoidal fracture; (2) breaks in rocks due to faulting or folding.

GEOMAGNETIC FIELD - the magnetic field of the Earth.

GEOMORPHOLOGY - The study of landforms.

GLAUCONITE - Dark-green mineral - a hydrous silicate of iron and potassium - occurring locally in the Cretaceous rocks and taken to indicate a slow rate of marine deposition.

GONDWANALAND - theoretical super-continent comprising approximately the present continents of the southern hemisphere, considered to have fragmented and drifted apart from Permian times onwards.

GRANITE - Coarse-textured acid igneous rock, essentially made up of quartz, feldspar and mica.

GNEISS - A coarse-grained regional metamorphic rock that shows compositional banding of minerals.

HARDGROUND - Levels within the Cretaceous Chalk in Antrim where due to breaks in the rate of deposition of deep sea mud or ooze, the sea bed became hardened and mineralized, usually showing green staining from the mineral glauconite.

HOLOCENE - In stratigraphy the younger or recent part of the Quaternary Period.

HORNFELS - Tough, fine-grained rock produced as a result of the thermal metamorphism of sediments by the heat from an igneous intrusion.

HYALOCLASTITE - Fragments of glassy material formed by the chilling or quenching of basaltic magma in water during subaqueous eruptions.

IAPETUS OCEAN - Earlier version of the Atlantic ocean that existed between 600 million years and 400 million years ago.

IGNEOUS ROCK - Rock formed by the crystallisation of molten material.

INTER-STADIAL - Short period of interglacial time during which the climate was warmer than the glacial or stadial phases.

INTRUSIVE ROCK - A body of molten igneous rock that has intruded or invaded older rocks beneath the surface, and subsequently solidified.

JURASSIC - Period of the stratigraphic column after the Trias and before the Cretaceous, from approximately 210 million years to 145 million years ago.

LAMELLIBRANCHS - Any member of the Pelecypoda or bi-valve molluscs, e.g. mussels.

LATERITE Weathering product, particularly of igneous rocks, under wet tropical conditions. Kaolinisation is followed by removal in solution of silica and other elements, leaving a characteristically red mixture of iron and aluminium oxides.

LAURASIA - Theoretical supercontinent comprising approximately the present continents of the Northern hemisphere, considered to have fragmented and drifted apart from Permian times onwards.

LAVA - Extruded igneous rock, commonly in the form of flat sheets (lava flows).

LIAS - In stratigraphy, part of the lower Jurassic Period. Characterised in northeast Ireland by grey clays and mudstones.

LIGNITE - A low-rank coal, dark brown in colour, formed by the partial decomposition of vegetable matter under anaerobic conditions.

LIMESTONE - A sedimentary rock made up largely of the calcium-carbonate mineral, calcite.

LITHOSPHERE - The outer rigid shell of the Earth, situated above the asthenosphere and comprising the crust and the upper brittle part of the mantle.

LODESTONE - The iron ore magnetite possessing polarity like a magnetic needle.

LOWER BASALTS - The Lower Basalt Formation of the Antrim Lava Group.

MAGMA - Molten rock material that forms igneous rocks on cooling. Magma that reaches the surface is extrusive and forms lava, magma that solidifies below the surface forms intrusive igneous rocks.

MAGNETIC NORTH POLE - The point where the earth's surface intersects the axis or line of the dipole that approximates the earth's magnetic field and where this field dips vertically downward.

MAGNETITE - Magnetic iron ore.

MANTLE - That region of the Earth between the crust and core.

MANTLE CONVECTION - The system of convection cells within the mantle thought to be the principal driving force for plate tectonics

MANTLE PLUMES - rising jets of hot, partially molten material from within the mantle and thought to be responsible for volcanism occurring away from plate boundaries such as at Hawaii.

MEGALITH - A Neolithic monument constructed from large stones.

MESOLITHIC - The "Middle Stone Age". In Ireland from about 6000-3000 BC.

METAMORPHISM - Change in the character of a rock due to heat and/or pressure, with the development of new minerals and structures.

MIDLANDIAN - Cold Stage in the Quaternary recognized in Ireland, from about
70 000 years to 10 000 years before present.

MID-OCEAN RIDGE - A major elevated linear feature of the sea floor. A characteristic type of constructive plate boundary where plates are diverging and new oceanic crust is being created.

MISFIT STREAM - Term used to describe a river which appears once to have been larger than it is today.

MORAINE - A glacial deposit of till left at the margins of an ice-sheet or glacier

MUDSTONE - A sedimentary rock made of clay-size particles; structureless and unlaminated.

NAUTILOIDS - Group of fossil sea creatures with an external chambered shell, related to modern squids.

NEOLITHIC - The "New Stone Age". In Ireland from 3000 to 1750 BC.

OLD RED SANDSTONE - Continental sandstones of the Devonian Period.

OLIGOCENE - In stratigraphy, a sub-division of the Tertiary Period.

OOZE - A fine-grained soft deep-sea mud with a high organic content.

PALAEOMAGNETIC POLE - The apparent position of the magnetic pole indicated by the contained magnetism of a rock. For many old rocks the palaeomagnetic pole does not coincide with the present magnetic pole and this is generally considered as evidence that the rock and its continental host have moved since the rock was formed.

PALAEOMAGNETISM - The science of the reconstruction of the earth's ancient magnetic field and from that the former positions of the continents using the evidence of magnetism retained in ancient igneous or metamorphic rocks as they cooled.

PALAGONITE - a yellow or orange coloured mineral formed by the alteration of basaltic glass and commonly associated with hyaloclastite breccias.

PANGAEA - A supercontinent that was formed of all the present continents and which existed about 200 million years ago.

PERIOD - In stratigraphy a major world-wide geological time unit.

PERMIAN - Period of the Stratigraphic Column after the Carboniferous and before the Trias, from approximately 280 million years ago to 245 million years ago.

PILLOW LAVAS - A type of lava extruded underwater in which many small pillow shaped tongues break through the chilled surface and quickly solidify.

PLANKTON - Floating organisms in oceans or lakes.

PLATE - One of the dozen or more segments of the lithosphere that move independently over the surface of the Earth, meeting at convergent or destructive margins and separating at divergent or constructive margins.

PLATE TECTONICS - The theory and study of the formation movement and interaction of lithospheric plates on the earth's surface.

PLEISTOCENE - In stratigraphy the older subdivision of the Quaternary Period.

PLESIOSAURS - A fossil marine reptile.

PLIOCENE - In stratigraphy the latest of the sub-divisions of the Tertiary Period.
PLUG - Vertical cylinder of igneous rock formed by the solidification of molten rock in the throat of an extinct volcano. A volcanic neck.

PRE-CAMBRIAN - In stratigraphy all rocks formed before the Cambrian Period, i.e. older than approximately 500 million years.

QUARTZ - A common mineral in granitic rocks , because of its hardness and durability it is the main component of sands and sandstones.

QUATERNARY - Period of the stratigraphic column after the Tertiary, from approximately 2 million years ago to Recent.

RADIOACTIVE DECAY - The breakdown of certain heavy elements to lighter elements by the emission of charged particles or radiation.

RADIOMETRIC DATING - The method of obtaining ages of geological materials by measuring the relative abundances of the radioactive element and the lighter element produced from it by radioactive decay.

RAISED BEACH - Coastal beaches and cliffs above the present shore line, cut at a time after the Ice Age when the sea level was higher due to melting of the ice caps.

RHYOLITE - Fine-grained acid lava. The extrusive equivalent of granite.

RIFTING - Faulting caused by divergence or tension in the crust.

SANDSTONE - Sedimentary rock composed of consolidated and cemented sand grains.

SCHIST - A metamorphosed rock, characterised by the development of secondary mica which gives it a lustrous appearance.

SEDIMENTARY ROCK - A rock formed by the accumulation and consolidation of mineral fragments transported by wind, water or ice to the deposition site, or formed by chemical action at the deposition site.

SHALE - A fine-grained sedimentary rock of clay or silt grade, with pronounced layering or lamination.

SILL - Flat-lying mass of igneous rock intruded conformably with the layering of the rocks into which it has been injected.

SILTSTONE - A fine-grained sedimentary rock of silt-grade particles, not conspicuously bedded.

SOLID MAP - A geological map showing the distribution of the solid or bedrock geology.

SPELEOTHEM - deposits in caves formed by calcium carbonate, the main constituent of limestone, being precipitated from dripping water, e.g. stalactites.

SPHEROIDAL WEATHERING - The formation of spherical boulders by chemical weathering along joints. Commonly found in weathered basalts.

STACK - A high rock off the coast, detached by marine erosion from the main cliff.

STALACTITES - a form of speleothem growing from a cave ceiling.

STRATIGRAPHY - The science of description, correlation and classification in stratified rocks, including the interpretation of the depositional environment of those rocks.

STRATIGRAPHIC COLUMN - The division of geological history into smaller units such as eras and periods through stratigraphy and the study of fossils.

SYNCLINE - A fold in rocks in which the beds dip inwards towards the centre. The opposite form is ANTICLINE.

TECTONICS - The study of the broad structural features of the Earth, and their causes.

TERTIARY - Period of the Stratigraphic Column after the Cretaceous and before the Quaternary, from approximately 65 million years to 2 million years ago.

THERMAL METAMORPHISM - See contact metamorphism.

THOLEIITE - Fine-grained conchoidal-fracturing igneous rock, rather less basic in composition than olivine-basalt, i.e. with a generally higher silica content.

TILL - An unconsolidated sediment consisting of an unsorted mixture of fragments of all sizes from clay to boulders, carried or deposited by glacial action.

TRIAS - Period of the Stratigraphic Column after the Permian and before the Jurassic, from approximately 250 million years to 210 million years ago.

TUFF - A rock formed by compaction of fragmental volcanic debris such as ash.

UNCONFORMITY - A surface of erosion or non-deposition separating younger rocks from older.

UPPER BASALTS - The Upper Basalt Formation of the Antrim Lava Group.

VENT - A volcanic pipe or opening, often filled with fragmental material.

VESICLE - Small gas-cavity, usually spherical, in a volcanic rock.

ZEOLITES - A group of minerals composed of the hydrated silicates of calcium and aluminium, sometimes with sodium and potassium. Commonly found in vesicles in the Antrim basalts.

APPENDIX 2.
FURTHER READING

Anglesea, M. and Preston, J. 1980. A philosophical landscape : Susanna Drury and the Giant's Causeway. Art History, vol.3(3), pp. 252-273.

Charlesworth, J.K. 1963. Historical geology of Ireland. Edinburgh, Oliver and Boyd, 565p.

Dietz, R.S. and Holden, J.C. 1972. The breakup of Pangaea. In Continents Adrift, Scientific American, W.H.Freeman.

Emeleus, C.H. and Preston, J. 1969. Field excursion guide : The Tertiary volcanic rocks of Ireland. Belfast, 70p.

Eyles, V.A. 1952. The composition and origin of the Antrim laterites and bauxites. Memoir of the Geological Survey of Northern Ireland.

Harland, W.B. 1969. Tectonic evolution of North Atlantic Region. In North Atlantic - Geology and Continental Drift. American Association Petroleum Geologists, Memoir 12, 817.

Holland, C.H.(Ed), 1981. A geology of Ireland. Edinburgh, Scottish Academic Press, 335p.

Lyle, P. 1980. A petrological and geo-chemical study of the Tertiary basaltic rocks of northeast Ireland. Journal of Earth Sciences, Royal Dublin Society, vol. 2(2), pp. 137-152.

Lyle, P. 1985. The petrogenesis of the Tertiary basaltic and intermediate lavas of northeast Ireland. Scottish Journal of Geology, vol. 21, pp. 71-84.

Lyle, P. and Preston, J. 1993. The geo-chemistry and volcanology of the Tertiary basalts of the Giant's Causeway area, Northern Ireland. Journal of the Geological Society, vol. 150, pp. 109-120.

P. T. R. S. 1694. Philosophical Transactions of the Royal Society, no. 212. July–August 1694.

Patterson, E.M. 1955. The Tertiary lava succession in the northern part of the Antrim plateau. Proceedings of the Royal Irish Academy, vol. 57, Section B, pp. 79-122.

Patterson, E.M. 1963. Tertiary vents in the northern part of the Antrim plateau, Ireland. Quarterly Journal of the Geological Society of London, vol. 119, pp. 419-443.

Patterson, E.M. and Swaine, D.J. 1955. A petrochemical study of Tertiary tholeiitic basalts : the Middle lavas of the Antrim plateau. Geochimica Cosmochimica Acta, vol. 8, pp. 173-181.

Preston, J. 1981. Tertiary igneous activity. In A geology of Ireland, ed C.H. Holland, Edinburgh, Scottish Academic Press, pp. 212-223.

Preston, J. 1982. Eruptive volcanism. In Igneous rocks of the British Isles, ed D.S. Sutherland, New York, Wiley, pp. 351-368.

Smith, A.J., Briden, J.C. and Drewry, G.E. 1973. Phanerozoic world maps. Special Paper Palaeontology 12, pp. 1-42.

Tomkeieff, S.I. 1940. The basalt lavas of the Giant's Causeway district of Northern Ireland. Bulletin Volcanologique, Serie 2, vol. 6, pp. 89-143.

Wilson, H.E. 1972. The regional geology of Northern Ireland. Belfast HMSO, 113p
.

Wilson, H.E. and Manning, P.I. 1978. Geology of the Causeway Coast : memoir for one-inch geological sheet 7, Northern Ireland Geological Survey, Belfast HMSO, vol. 1, 72p.

Wyllie, P.J. 1976. The way the Earth works : An introduction to the new global geology and its revolutionary development. John Wiley and Sons, 296p.

APPENDIX 3. SUMMARY OF EXCURSION LOCALITIES

LOCALITY NUMBER	LOCALITY NAME	FEATURES	CHAPTER IN TEXT	PAGE NUMBER	GRID REFERENCE
1	Portrush Sill	'Portrush Rock' Dolerite sill	5.1	44	C 857 414 Sheet 4
2	White Rocks Vent Series	Vent series in chalk	5.2	46	C 892 409 Sheet 4
3	Craignahulliar Quarry		5.3	49	C 883 389 Sheet 4
3a	North-western exposure	Laterite, lignite amygdales	5.3	50	
3b	Central exposure and lava blisters	hyaloclastite, sediments	5.3	51	
3c	North-eastern side quarry	colonnade with chisel marks	5.3	52	
4	Giants Causeway section		5.4	52	
4a	To Great Stookan	Laterite	5.4	54	C 944 443 Sheet 4
4b	Windy Gap and Great Stookan basalts	Spheroidal weathering, amygdaloidal	5.4	54	C 943 446 Sheet 4
4c	Giants Causeway	The Causeways	5.4	55	C 946 448 Sheet 4

LOCALITY NUMBER	LOCALITY NAME	FEATURES	CHAPTER IN TEXT	PAGE NUMBER	GRID REFERENCE
4d	Giants Organ, Port Noffer	colonnade, entablature	5.4	57	C 952 448 Sheet 4
4e	Port Noffer to Port Reostan	Laterite, 'Giants Eye', dyke	5.4	58	C 951 450 Sheet 4
4f	Port Reostan	colonnade, 'Harp', 'Chimney'	5.4	58	C 951 452 Sheet 4
5	White Park Bay to Ballintoy Harbour		5.5	58	
5a	Ballintoy Harbour car park	Fault breccia, cave	5.5	60	D 037 455 Sheet 5
5b	Coastal path : Tertiary	Agglomerate, lavas	5.5	60	D 036 456 Sheet 5
5c	Coastal path : raised beach	sea stack, sea arch	5.5	61	D 035 454 Sheet 5
5d	White Park Bay	Lias, Cretaceous, landslips	5.5	62	D029 449 Sheets
5e	Bendoo plug	dolerite, metamorphism	5.5	63	D 044 455 Sheet 5
6	Carrickarade	sill, volcanic vent	5.6	64	D 062 449 Sheet 5
7	Kinbane Head	volcanic vent	5.7	67	D 088 438 Sheet 5

Sheet numbers refer to 1:50000 Ordnance Survey Discoverer maps